Cultural Mega-Events and Heritage

This book analyses case studies of heritage-rich cities that hosted mega-events to discuss emerging challenges, controversies, and accomplishments.

The future of mega-events has never been more uncertain. The ongoing COVID-19 pandemic has introduced an unparalleled level of doubt regarding the kind of mega-events that will take place in the coming years. This book arrives at a quite unique moment of reflection. Prior to 2020, cities were already questioning the traditional format of mega-events (e.g. Olympics and Expo) while other cultural mega-events have been spreading and gaining popularity, thanks in part to typically lower costs of infrastructures and venues, far more adaptable arrangements, spatial distribution and time frame for hosting. In these ways, they have already been demonstrating higher flexibility in which to respond to future health and safety constraints. When it comes to the relation to the existing city, cultural mega-events have been planned, implemented, and studied far more than any other. By leveraging the richness of cultural mega-events, this multidisciplinary collection deepens the intersection between events and cultural heritage. The chapters in this book provide a new theoretical framework, critical questions, and relevant case studies to argue that the nexus between mega-events and heritage is a key challenge for many cities in Europe and beyond.

The chapters in this book were originally published as a special issue of *European Planning Studies.*

Davide Ponzini is Full Professor of Urban Planning at Politecnico di Milano, Italy. He has been Visiting Scholar at Yale, Columbia University, TU Munich, and other universities. His research activity focuses on planning theory, urban and cultural policy, and contemporary architecture. His latest book is titled: *Transnational Architecture and Urbanism* (Routledge 2020).

Cultural Mega-Events and Heritage

Challenges for European Cities

Edited by

Davide Ponzini

Routledge
Taylor & Francis Group
LONDON AND NEW YORK

First published 2023
by Routledge
4 Park Square, Milton Park, Abingdon, Oxon, OX14 4RN

and by Routledge
605 Third Avenue, New York, NY 10158

Routledge is an imprint of the Taylor & Francis Group, an informa business

British Library Cataloguing-in-Publication Data
A catalogue record for this book is available from the British Library

ISBN13: 978-1-032-44583-0 (hbk)
ISBN13: 978-1-032-44584-7 (pbk)
ISBN13: 978-1-003-37289-9 (ebk)

DOI: 10.4324/9781003372899

Typeset in Minion Pro
by codeMantra

Publisher's Note
The publisher accepts responsibility for any inconsistencies that may have arisen during the conversion of this book from journal articles to book chapters, namely the inclusion of journal terminology.

Contents

Citation Information

The chapters in this book were originally published in the journal *European Planning Studies*, volume 30, issue 3 (2022). When citing this material, please use the original page numbering for each article, as follows:

Introduction

Introduction to the special issue – cultural mega-events and heritage: challenges for European cities
Davide Ponzini
European Planning Studies, volume 30, issue 3 (2022) pp. 427–436

Chapter 1

Short-term gains and long-term challenges to learning from mega-event planning in the city of Genoa
Zachary Mark Jones
European Planning Studies, volume 30, issue 3 (2022) pp. 437–456

Chapter 2

A mega-event in a small city: community participation, heritage and scale in the case of Pafos 2017 European Capital of Culture
Evanthia Dova, Angeliki Sivitanidou, Natia R. Anastasi and Julia Georgi-Nerantzia Tzortzi
European Planning Studies, volume 30, issue 3 (2022) pp. 457–477

Chapter 3

A heritage-inspired cultural mega-event in a stigmatized city: Hull UK City of Culture 2017
Enrico Tommarchi and Franco Bianchini
European Planning Studies, volume 30, issue 3 (2022) pp. 478–498

Chapter 4

Cultural events and heritage policy for the Milan Expo 2015: experimental intersections between mega-event and city
Stefano Di Vita
European Planning Studies, volume 30, issue 3 (2022) pp. 499–513

Chapter 5

Culture and heritage as a means to foster quality of life? The case of Wrocław European Capital of Culture 2016
Joanna Sanetra-Szeliga
European Planning Studies, volume 30, issue 3 (2022) pp. 514–533

Chapter 6

Liverpool's European Capital of Culture legacy narrative: a selective heritage?
Tamara West
European Planning Studies, volume 30, issue 3 (2022) pp. 534–553

Chapter 7

Internationalizing small-sized cities through mega-events: the case of Matera-Basilicata 2019 European Capital of Culture
Marina Rotolo
European Planning Studies, volume 30, issue 3 (2022) pp. 554–565

Chapter 8

Heritage and cultural mega-events: backgrounds, approaches and challenges
Jacek Purchla
European Planning Studies, volume 30, issue 3 (2022) pp. 566–572

Notes on Contributors

Natia R. Anastasi, School of Architecture, Engineering, Land and Environmental Sciences, Neapolis University, Pafos, Cyprus.

Franco Bianchini, Centre for Cultural Value, Leeds, UK.

Stefano Di Vita, Department of Architecture and Urban Studies, Politecnico di Milano, Italy.

Evanthia Dova, School of Architecture, Engineering, Land and Environmental Sciences, Neapolis University, Pafos, Cyprus.

Zachary Mark Jones, Department of Architecture and Urban Studies, Politecnico di Milano, Italy.

Davide Ponzini, Department of Architecture and Urban Studies, Politecnico di Milano, Italy.

Jacek Purchla, Department of Economic and Social History, Krakow University of Economics, Poland.

Marina Rotolo, École Nationale Supérieure d'Architecture de Bretagne, Rennes, France.

Joanna Sanetra-Szeliga, Department of Public Economy and Administration, Cracow University of Economics, Poland.

Angeliki Sivitanidou, School of Architecture, Engineering, Land and Environmental Sciences, Neapolis University, Pafos, Cyprus.

Enrico Tommarchi, School of Law and Social Sciences, London South Bank University, UK.

Julia Georgi-Nerantzia Tzortzi, Department of Architecture, Built Environment and Construction Engineering, Politecnico di Milano, Italy.

Tamara West, Department of Languages, Cultures and Film, University of Liverpool, UK.

Introduction – Cultural mega-events and heritage: challenges for European cities

Davide Ponzini

ABSTRACT
The future of mega-events has never been more uncertain. The ongoing COVID-19 pandemic has introduced an unparalleled level of doubt regarding the kind of mega-events that will take place in the coming years. While the 2019 call for papers could not predict such a radical shock, this special issue arrives at a quite unique moment of reflection. Prior to 2020, cities were already questioning the traditional format of mega-events (e.g. Olympics and Expo) while other cultural mega-events have been spreading and gaining popularity, thanks in part to typically lower costs of infrastructures and venues, far more adaptable arrangements, spatial distribution and time frame for hosting. In these ways, they have already been demonstrating higher flexibility in which to respond to future health and safety constraints. When it comes to the relation to the existing city, cultural mega-events have been planned, implemented and studied far more than any other. By leveraging the richness of cultural mega-events, this special issue deepens the intersection between events and cultural heritage in particular. This introduction provides a theoretical framework that cuts across mega-event and heritage research fields. It introduces the questions and approaches of the following contributions, arguing that the nexus between mega-events and heritage is a key challenge for many cities in Europe and beyond.

Introduction: cultural mega-events and heritage

Even before the COVID-19 pandemic, criticism toward sport and expo mega-events was on the rise, pushing several cities to rethink the traditional mega-event rationale. From multiple sides (including environmental, civic and public finance concerns), the typical boosterist approach has been under attack as large infrastructures and new facilities, which promise a stronger global image for the city and consequent increase in tourism, often fail to deliver (Gold and Gold 2008; Kassens-Noor 2012; Müller 2015a; Zimbalist 2015; Flyvbjerg, Budzier, and Lunn 2020). In terms of urban planning, experts and scholars alike have come to criticize the creation of brand new self-standing platforms in city outskirts that have become common in past decades (Di Vita and Morandi 2018). Significant and systematic withdrawals of cities from bids to host the

Summer and Winter Olympics Games (e.g. Boston, Rome, Hamburg and Budapest – see Lauermann and Vogelpohl 2017) have accompanied more widespread sensitivity toward sustainable and long-term intervention – both motivated by serious and widely-acknowledged financial, planning and political problems. These developments led the International Olympic Committee (IOC) towards encouraging smaller budgets, temporary structures and the re-use of pre-existing infrastructures and facilities to be adapted not only for the mega-event itself but for the future of the city. Meanwhile, FIFA has also expanded the traditional hosting model. For example, three countries in North America will share the 2026 World Cup to reduce the expenditure on any single host country. These attempts at making, particularly sport, mega-events more diffused, acceptable and sustainable will inevitably face the complexities of planning and delivering such events in the context of historic, modern and contemporary cities. These new approaches increasingly rely on existing facilities that are typically located within more lived-in urban environments. Controversies may particularly arise in European cities, especially when one considers the important role of both tangible and intangible heritage in terms of land-use planning, political consensus building and city image, to mention a few. However, cultural mega-events have remained or even grown in popularity in recent years, in large part due to their reduced costs and regular re-use of existing venues (Di Vita and Wilson 2020 provide an interesting insight into smaller events).

This nexus between heritage and mega-events might be a double-edged sword, one clearly dependent upon specific planning contexts and processes. On one hand, mega-event planning dynamics may streamline resources and new projects for heritage policies, places and facilities. On the other hand, the strategy of hosting a mega-event and related goals (e.g. attracting high volumes of visitors for short periods, with the intention to boost long-term tourism trends – see Liu 2014) can commodify and irreversibly affect the physical qualities and social uses of an area and its heritage value. Despite the opportunities and threats that mega-events tend to generate in heritage-rich cities and the importance of related planning matters, limited research is available (Jones and Ponzini 2018). Also, as events come to adapt to COVID-19 restrictions, the spreading out of event spaces and locations is becoming a prominently proposed solution which may also suggest a greater interaction between historic/heritage spaces in cities and mega-events.

The contributions in this *European Planning Studies* special issue looked at specific mega-events that have been dealing with similar problems. The attention of the authors focuses on one or more of the following aspects:

- the use and transformation of historic sites and urban landscapes (including tangible and intangible heritage, as well as the city image) for/during the mega-event, and their longer-term implications and effects;
- the response of heritage policy making to the planning of mega-events that – due to necessity – interacted with heritage actors and measures, projects or ways of operating;
- the short-and-medium-term implications for city planning, governance and politics in terms of participation (e.g. involvement of local communities at different stages, empowering of tourism-and-event constituencies, political lock-ins and the continuing support to the local event industry);

- the discovery of or focussing attention on less obvious heritage, narratives and sites in the city or in relation to the mega-event and for the more general goal of boosting media exposure and tourism;
- mega-event policy discourses, rationales, planning tools and impacts are discussed in relation to heritage and heritage policies and vice versa.

All papers focus on an in-depth case-study but touch upon a range of issues and often explicitly refer to other relevant examples. The contributions cover cities across all quadrants of Europe and the ways in which one or more mega-event affected their development in the medium to long-term, primarily concerning their heritage. In some cases, specific plans and projects at the core of the mega-event proved to be relevant to heritage matters, in others the momentum and substantial investments in terms of infrastructure and complementary events surrounding the mega-event itself affected heritage the most. Given the variety of experiences in Europe, it is not possible to expect overarching and generalizable theories, while the diversity of such experiences allows multiple lessons for future host cities to learn from. The idea underpinning this collection of articles is that academic research can help inform the international debate and anticipate some of the emerging urban problems and solutions (Ponzini 2016) and that, in particular, one can learn across different types of mega-events, without the need for one-on-one comparisons or generalizable theories (Ponzini 2020). This introductory paper provides a brief theoretical and methodological framework regarding the crossroads between mega-event studies and heritage policy studies, outlining the key concepts and main questions that the contributions address.

Considering cultural mega-events

This special issue 'Cultural Mega-events and Heritage: Challenges for European Cities' brings together international scholars to discuss the diverse experiences of mega-events such as the city of culture initiatives (European Capital of Culture – ECoC – and the UK City of Culture) and Expo, paying close attention to the relationship with tangible and intangible heritage. As the work of Jones and myself showed (Jones and Ponzini 2018; Jones 2020), this area of overlap between mega-events and heritage policy has been clearly understudied in literature and there is now a great opportunity for Europe, in particular, to reflect upon the potentiality of cultural mega-events. With over 35 years of continuous planning and experimentation in over 60 cities in Europe, the ECoC programme has fostered significant changes not only in terms of cultural policy but also in urban facilities and city image. It has touched different cities at the heart of as well as across all corners of Europe, from global capitals to medium-size and smaller cities, welcoming from hundreds of thousands to millions of visitors during the year of celebration. In many cases, the programme pushed host cities to repurpose, re-use and rethink their existing cultural facilities, cultural places and public realm altogether while generating a new future for these assets as part of the city. These and other reasons make the ECoC an interesting laboratory for observing the relationship between mega-events and cultural heritage more generally, as well as an important testing ground for the research and planning of mega-events to consider. In fact, the ECoC has become a reference model for a number of national and international programmes with varying features, e.g. the Ibero-American Capital of Culture, the Culture City of East Asia, the UK City of Culture, Italian Capital of Culture (Ponzini et al. 2019). These events further expand the experimentation and situations where cultural mega-events interact with cities and heritage, thus providing a quite large untapped field of observation for both research and policy learning.

The Expo has a much longer history and perhaps requires a broader cultural acceptance than the ECoC as it includes technical achievements as well as more general innovations with typically a more stringent commercial orientation. Its cultural dimension remains relevant, but perhaps the Expo's relationship with the city and its existing fabric and heritage are less intense than the ECoC, due to the size, functional and security standard requirements of its site. On many occasions, Expo sites were self-standing platforms that became the opportunity for cities to drive significant expansions or additions of their structure, or radical transformation of brownfields through a 'tabula rasa' approach, rather than the renovation of the existing fabric. The length of Expos can reach six months and the budgets vary significantly according to the need for infrastructure. It is undeniable that there are significant differences among the manifestations of these cultural events as compared to sports events like the Olympic Games.

Despite the fact that sport has a cultural component (as argued in the special issue edited by Gammon, Ramshaw, and Waterton 2013) and although sport mega-events are often accompanied by a rich calendar of cultural events that amplifies the media exposure and other expected effects (see Garcia 2012; Low and Hall 2012), this special issue keeps 'sport' mega-events separate from 'cultural' mega-events. It is worth reflecting briefly on existing literature in order to discuss how and to what extent the common 'mega-event' label may help mutual learning across types of events, especially when discussing their relationship with cities and heritage.

Mega-events beyond measures

The international academic debate generated a high number of definitions, distinctions and arguments regarding mega-events, large events, hallmark events, festivals and others (Hall 1989; Horne 2007; Müller 2015b). In many cases, the discussion concentrates on quantitative thresholds that help in classifying events by the needed budget and actual costs, duration, number of visitors and the like. Urban scholars expanded the observation toward city impact (Garcia, Melville, and Cox 2010; Müller 2015b) and suggested ways to measure the effects. Surely the hundreds of millions of euros spent for a cultural mega-event cannot directly compare to the multi-billions needed to host a world-class mega-event such as the Olympics or the FIFA World Cup. Also, the six-month celebration for the Expo or the year-long celebrations of an ECoC varies drastically from the few weeks dedicated to the Olympics/Paralympics. However, when one focuses their attention on the range of urban effects and broader changes induced, separating types of events by quantitative figures may end up preventing a certain level of learning permeability across cities and events. For this reason, there are widely used and more flexible definitions of mega-events to refer to when discussing the matters in this special issue (Jones 2020). Among them, the one proposed by Roche (2000, 1) '... large-scale cultural (including commercial and sporting) events which have a dramatic character, mass popular appeal and international significance' seems specific and at the same time flexible enough to include sport and cultural mega-events. In his discussion of cultural mega-events, Jones (2020) argues that the boundaries between different types of mega-events have become more and more blurred, even when one looks at specific ways of measuring mega-events (e.g. attendance, public and private expenditure).

In addition, considering the proportion of the event and its urban effects in relation to the size and functioning of the host city is crucial. European cities hosting large events span from global capitals like London or Paris to smaller cities of 50,000 inhabitants or less, which have greatly reduced media visibility, economic structures and opportunities more generally. In this sense, the quantitative requirements to qualify as being 'mega' may be discussed in terms of its meanings, expectations, impact and legacies depending upon the context. While recognizing the substantially different situations, one may see how programmes such as the ECoC or the Expo, with their different budgets or attendance, may yet ultimately result in paramount urban changes in relation to the size and operating of the host city. The importance of both tangible and intangible effects within European cities make cultural mega-events interesting and relevant as experiences to observe (Richards and Palmer 2010; Ponzini et al. 2020), despite the fact that the comparison with other types of mega-events or another geographical context may not be linear or simple.

European cities, mega-events and heritage

In this special issue, cultural heritage (as opposed to natural) and built heritage form the core, with several articulations in terms of historic areas, heritage places and urban landscapes (Jones and Ponzini 2018; Ponzini et al. 2019). Intangible heritage is also important in our discussion, as the narration and processes of development and strengthening the city image involve symbols, customs and imaginaries that buildings and urban spaces do not necessarily embody. There are different ways of interpreting the importance of heritage depending on the national and regional planning systems, different agencies and measures or approaches.

Firstly, the explorations in this special issue approach the case-study city without expecting it to be the manifestation of any single and general model of a 'European city'. It is even less expected that the complex relations between mega-events and heritage will necessarily work in identical manners across cities, given the differences in terms of institutional and planning systems, heritage policy and powers. However, there are a number of similarities to consider, such as the political models and democratic governance applications in different cities of Europe, when it comes to important political decisions like the bidding for a mega-event. Yet one cannot expect a duplication of processes or their outcomes. In some cases, the mayor decides, in others, a referendum or other ways of gauging public opinion are required to support decisions at critical turns. The power is given to heritage authorities and even their existence and institutional location also vary across Europe (centralized at the national level, articulated at the regional level, present as part of the local government or eventually heard as an autonomous nongovernmental body). Despite these differences, there remains a common understanding of the importance of the past in making planning decisions for the future, at the bare minimum in terms of tourism exploitation of given sites and identity-bearing objects and values.

Secondly, in connection to the previous point, one must recognize that the EU has effectively worked to reassert certain political values centred on democratic governance within and outside its member states. Since the 1990s, the spreading of urban policy ideas and governance models – championed by pilot projects, Structure Funds, policy recommendations and white papers – led planning scholars to discuss the Europeanisation of planning (Dühr, Stead, and Zonneveld 2007; Colomb, Dühr, and Nadin 2010). The ECoC itself can be considered as a motor of this partial convergence in terms of urban planning and cultural policy making as it has been one of the many ways in which the circulation of international expertise within Europe was made easier by means of competition. More generally, the introduction of common standards for services (e.g. the 1995 General Agreement on Trade in Services and the 2006 Bolkestein Directive) and the creation of a large set of intra-European networks of experts and decision-makers for common learning and best-practice transfer contributed to this process. The existence of common values regarding heritage and the European Landscape Convention by the Council of Europe started to build bridges among quite different situations since the early 2000s.

Thirdly, the richly layered physical and symbolic presence and political relevance of heritage are particularly relevant in European cities, making the interventions for and delivery of mega-events typically more complex within the historic city rather than in a blank-slate setting. The potential frictions and disruptive effects seem amplified by the presence of the past in given areas of European cities. In this sense, there is not any single 'European city' or European city model to be searched for in this special issue. The intention is to use the common features as well as the variety of cases to discuss critical issues that are experienced when planning mega-events in heritage-rich cities in Europe and beyond. Other regional explorations of similar as well as different mega-events have also been carried out, to which this special issue builds upon (see, among others: Sykes 2011; Müller and Pickles 2015; Ponzini and Jones 2015)

Making the case about cultural mega-events and heritage

The articles in this special issue discuss different problems regarding the mega-event/heritage nexus that has the potential to become an increasingly important issue for European cities. By using the in-depth analysis of cases and more descriptive examples, the articles allow for informed and research-heavy views of different local experiences, rather than composing the building blocks for a systematic comparison or a general theory. Several topics in this introduction intertwine and find their own discussion in the following articles.

In his 'Short-term Gains and Long-term Challenges to Learning from Mega-event Planning in the City of Genoa', Zachary M. Jones presents the long-term strategy of the city of Genoa that combined mega-events and the regeneration of one of the most extensive historic city centres in Europe to re-activate its de-industrialized yet, at that time, publicly inaccessible waterfront. Within 12 years, Genoa hosted three large events (1992 Expo, 2001 G8 Summit and 2004 ECoC) and experimented with new governance styles and arrangements, involving the cultural policy sector and heritage preservation in particular. The achievement of the title of UNESCO World Heritage Site confirms the results of this innovative period, but at the same time marked a turning point where subsequent stages of planning rapidly lost the earlier momentum encouraged by the mega-events.

The strategic importance of the ECoC for spatial planning and for reactivating heritage places and policies is documented in the article 'A Mega-event in a Small City: Community Participation, Heritage and Scale in the Case of Pafos 2017 European Capital of Culture' by Evanthia Dova, Angeliki Sivitanidou, Natia Anastasi and Julia Georgi. Clearly, the small size of the Cypriot host city amplified the dynamics and effects of the international event, in terms of urban space and local involvement. In all phases of bidding, event planning, implementation and the post-event, urban heritage played a crucial role in bringing together different actors to give new life to cultural and public places while addressing dissonant or long-silenced stories.

In their article 'A Heritage-inspired Cultural Mega-event in a Stigmatised City: Hull UK City of Culture 2017', Enrico Tommarchi and Franco Bianchini explore how a cultural mega-event mobilized built heritage to process the difficult past of a medium-sized city in Northern England. Authorities implemented different strategies to engage visitors and citizens in the historic urban environment through collective learning experiments, the use of art installations and performances as well as the rediscovery and spectacularisation of heritage. The positive effects of the mega-event however delivered only part of the expected changes to the city.

Stefano Di Vita explores the role of cultural programmes that complemented the Expo while linking to heritage. In 'Cultural Events and Heritage Policy for the Milan Expo 2015: Experimental Intersections between Mega-event and City' he argues how the combination of the mega-event and a coordinated calendar of smaller events benefited the image and tourism attractiveness of the city over the long term. Despite the limits in terms of spatial planning vision, the availability of additional resources for the city and the hard deadline of 2015 unlocked previously blocked heritage projects. The smaller events and their widespread locations allowed for a synergetic and regenerative interaction between heritage buildings and sites in the city centre and beyond.

Joanna Sanetra-Szeliga's article 'Culture and Heritage as a Means to Foster Quality of Life? The Case of Wrocław European Capital of Culture 2016' connects the mega-event planning with projects and initiatives levering urban heritage. Wrocław's history and approach to mega-event planning implied significant experimentation with participation and different forms of tangible and intangible heritage. The mega-event marked an important moment in the recognition of given heritage places and developing a sense of belonging at multiple scales, from the urban to the very local.

Tamara West questions the use of cultural programmes, heritage and the building of city image in 'Liverpool's European Capital of Culture Legacy Narrative: A Selective Heritage?'. The connection between heritage and economic development affected the planning and delivering of the ECoC as well as subsequent phases. In West's view, this approach short-circuited the imminent threat of delisting of the city's UNESCO World Heritage Site after the event. The article calls for further attention to the narratives and unintended consequences of widespread pro-growth approaches that permeate mega-event politics and planning.

Marina Rotolo's 'Internationalising Small-sized Cities through Mega-events: The Case of Matera-Basilicata 2019 European Capital of Culture' explores the controversial links between the ECoC and the fragile urban heritage of Matera, one of the oldest continuously inhabited cities in the world (also a UNESCO World Heritage Site). The transformation of the city image and the difficulties of transforming its environment become increasingly evident in the different phases of the mega-event.

In his commentary, Jacek Purchla problematizes the conception and role of heritage and of the urban landscape in historic European cities. By discussing the trajectory of the European Capital of Culture programme, his article brings to the fore challenges regarding the accelerating and increasing need for heritage and subsequent expanding demand from multiple socio-economic interests at play in contemporary cities and more broadly.

The original call for papers back in 2019 could not have anticipated the dramatic global changes that would soon occur and particularly their impacts on mega-events and cultural policy. In 2021, it remains difficult to predict how long and in-depth the pandemic will continue to affect cities and events. It is however undeniable that European cities hosting a mega-event will have to deal with many of the problems discussed in this special issue, starting from the future European Capitals of Culture like Kaunas 2022, Veszprém 2023, as well as the Olympics to be held in heritage-rich cities like Paris in 2024 or Milan in 2026. Moving from an academic perspective towards the policy realm, the *Charter for Mega-events in Heritage-rich Cities*[1] provides a set of policy principles and recommendations to assist decision-makers preparing for events at different levels. The papers in this special issue can contribute to the broader research and policy discussion of the future of mega-events and of their interaction with cultural heritage, particularly within European cities.

Note

1. The full text of the Charter can be downloaded here https://mck.krakow.pl/homee-charter.

Acknowledgments

The core ideas for this special issue derived from long-term research conducted in collaboration with Zachary M. Jones at the Department of Architecture and Urban Studies, Politecnico di Milano. In particular, sections of this paper are based on the research conducted within the JPI Cultural Heritage research project 'HOMEE – Heritage Opportunities/threats within Mega-Events in Europe' (with a grant issued by the Italian Ministry of Education, University and Research) that explored similar problems and involved a number of the contributors to this issue. An earlier version of this introductory article was commented by Stefano Di Vita, Evanthia Dova, Zachary M. Jones, Jacek Purchla, Marina Rotolo, Angeliki Sivitanidou, Enrico Tommarchi and Tamara West whom the author wants to thank for their substantial input. The usual caveat applies.

Disclosure statement

No potential conflict of interest was reported by the author(s).

References

Colomb, C., S. Dühr, and V. Nadin. 2010. *European Spatial Planning and Territorial Cooperation*. London: Routledge.

Di Vita, S., and C. Morandi. 2018. *Mega-events and Legacies in Post-metropolitan Spaces. Expos and Urban Agendas*. Cham: Palgrave MacMillan.

Di Vita, S., and M. Wilson. 2020. *Planning and Managing Smaller Events: Downsizing the Urban Spectacle*. London: Routledge. doi: 10.4324/9780429266263
Dühr, S., D. Stead, and W. Zonneveld. 2007. "The Europeanization of Spatial Planning Through Territorial Cooperation." *Planning Practice and Research* 22 (3): 291–307. doi:10.1080/02697450701688245
Flyvbjerg, B., A. Budzier, and D. Lunn. 2020. "Regression to the Tail: Why The Olympics Blow up." *Environment and Planning A: Economy and Space* 52 (2): 233–260. doi:10.1177/0308518X20958724
Gammon, S., G. Ramshaw, and E. Waterton. 2013. "Examining the Olympics: Heritage, Identity and Performance." *International Journal of Heritage Studies* 19 (2): 119–124. doi:10.1080/13527258.2012.687395
Garcia, B. 2012. *The Olympic Games and Cultural Policy*. London: Routledge.
Garcia, B., R. Melville, and T. Cox. 2010. *Creating an Impact: Liverpool's Experience as European Capital of Culture*. Liverpool: Impacts.
García, B., and T. Cox. 2013. *European Capitals of Culture: Success Strategies and Long-term Effects*. Luxembourg: European Parliament Committee on Culture and Education.
Gold, J., and M. Gold. 2008. "Olympic Cities: Regeneration, City Rebranding and Changing Urban Agendas." *Geography Compass* 2 (1): 300–318. doi:10.1111/j.1749-8198.2007.00080.x
Hall, C. 1989. "The Definition and Analysis of Hallmark Tourist Events." *GeoJournal* 19 (3): 263–268. doi:10.1007/BF00454570
Horne, J. 2007. "The Four 'Knowns' of Sports Mega-events." *Leisure Studies* 26 (1): 81–96. doi:10.1080/02614360500504628
Jones, Z. 2020. *Cultural Mega-events: Opportunities and Risks for Heritage Cities*. London: Routledge.
Jones, Z., and D. Ponzini. 2018. "Mega-events and the Preservation of Urban Heritage: Literature Gaps, Potential Overlaps and a Call for Further Research." *Journal of Planning Literature* 33 (4): 433–450. doi:10.1177/0885412218779603
Kassens-Noor, E. 2012. *Planning Olympic Legacies: Transport Dreams and Urban Realities*. New York, NY: Routledge.
Lauermann, J., and A. Vogelpohl. 2017. "Fragile Growth Coalitions or Powerful Contestations? Cancelled Olympic Bids In Boston and Hamburg." *Environment and Planning A: Economy and Space* 49 (8): 1887–1904. doi:10.1177/0308518X17711447
Liu, Y. 2014. "Cultural Events and Cultural Tourism Development: Lessons from the European Capitals of Culture." *European Planning Studies* 22 (3): 498–514. doi:10.1080/09654313.2012.752442
Low, D., and P. V. Hall. 2012. "The 2010 Cultural Olympiad: Playing for the Global or the Local Stage?" *International Journal of Cultural Policy* 18 (2): 131–150. doi:10.1080/10286632.2011.567332
Müller, M. 2015a. "The Mega-event Syndrome: Why So Much Goes Wrong in Mega-event Planning and What to Do About It." *Journal of the American Planning Association* 81 (1): 6–17. doi:10.1080/01944363.2015.1038292
Müller, M. 2015b. "What Makes an Event a Mega-event? Definitions and Sizes" *Leisure Studies* 34 (6): 627–642. doi:10.1080/02614367.2014.993333
Müller, M., and J. Pickles. 2015. "Global Games, Local Rules: Mega-Events in the Post-Socialist World." *European Urban and Regional Studies* 22 (2): 121–127. doi:10.1177/0969776414560866
Ponzini, D. 2016. "Introduction: Crisis and Renewal of Contemporary Urban Planning." *European Planning Studies* 24 (7): 1237–1245. doi:10.1080/09654313.2016.1168782
Ponzini, D. 2020. *Transnational Architecture and Urbanism: Rethinking How Cities Plan, Transform, and Learn*. London: Routledge.
Ponzini, D., F. Bianchini, J. Georgi-Nerantzia Tzortzi, and J. Sanetra–Szeliga. 2020. *Mega-events and Heritage: The Experience of Five European Cities*. Krakow: International Cultural Centre.
Ponzini, D., and Z. Jones. 2015. "Meanings, Urban Products and Forms of "European Cities" in the European Capital of Culture Program." *Urbanistica* 155: 70–78.

Ponzini, D., Z. Jones, F. Bianchini, E. Tommarchi, J. Georgi, E. Dova, A. Sivitanidou, et al. 2019. *HOMEE Literature Review of Mega – events Addressing Cultural Heritage Issues.* http://www.tau-lab.polimi.it/homee-literature-review-regarding-mega-events-cultural-heritage
Richards, G., and R. Palmer. 2010. *Eventful Cities: Cultural Management and Urban Revitalisation.* Oxford: Elsevier.
Roche, M. 2000. *Mega-events and Modernity.* London: Routledge.
Sykes, O. 2011. "Introduction: European Cities and Capitals of Culture – A Comparative Approach." *Town Planning Review* 82 (1): 1–12. doi:10.3828/tpr.2011.8
Tommarchi, E., L. Hansen, and F. Bianchini. 2018. "Problematising the Question of Participation in Capitals of Culture." *Participations* 15 (2): 154–169.
Zimbalist, A. 2015. *Circus Maximus: The Economic Gamble Behind Hosting the Olympics and the World Cup.* Washington, DC: Brookings Institution Press.

Short-term gains and long-term challenges to learning from mega-event planning in the city of Genoa

Zachary Mark Jones

ABSTRACT

For decades, a mega-event strategy to urban development has been popular with developers and politicians as a way to deliver fast results that can help to regenerate areas of cities and boost a city's global image. While this approach can deliver quick results in the short term, the literature has not yet investigated the lasting affects this has at an institutional level on city planning and decision-making processes over the long term. One interesting case to investigate this phenomenon is the city of Genoa, which hosted three large events over a period of 12 years to regenerate the city centre and reconnect it to its de-industrialized waterfront. While the 1992 Expo, 2001 G8 Summit and 2004 European Capital of Culture significantly altered the urban fabric of the city centre, this paper will take a closer look at the new approaches and strategies decision makers implemented in the planning of these events, analysing whether they continued to utilize these planning tools and approaches following the events. The paper finds that while the city improved in its planning of events through the process of hosting the three events, it struggled to maintain such innovative approaches as part of routine practice post-event.

1. The appeal of the mega-event strategy

Mega-events have long been popular albeit contentious strategies to significantly invest in new infrastructure, development and promote cities on a global stage (Duan, Yi, and Kunzmann 2016; Wu, Li, and Lin 2016). Though by no means always realized, event planners and promoters often intend to improve the city, both tangibly and intangibly, with benefits continuing for years after the event thanks to the city's newly founded or renewed global position. Several examples are commonly used to reinforce this narrative, particularly the 1992 Barcelona Olympic Games or more recently the 2012 London Olympic Games. The 'Barcelona Model' in particular has become a wide spread and diffused narrative detailing the opportunities provided to cities through hosting a mega-event (González 2011; Smith 2005). Yet city decision makers, academics and local citizens alike have become increasingly sceptical of the mega-event strategy due to examples such as the Athens 2004 or Sochi 2014 Olympics where the professed transformations and long-term benefits failed to materialize as promised (Gold 2017; Müller 2014).

While debates over the socio-economic and physical impacts of mega-events will surely continue, another key aspect of the mega-event strategy pertains to the changes to planning practice and policy that these processes typically introduce (Lauermann 2019; Silvestre 2009). Basso (2015, 2017) has described these processes as *exceptional routines* where the planning of mega-events tends to utilize existing emergency measures typically reserved for disaster response. These can include special legislation to provide extensive funding and bypass normal planning regulations to fast track decision making and implementation of projects. The management and governance of mega-events can also introduce significant changes from the formation of new coalitions of existing local stakeholders to the creation of entirely new agencies tasked with delivering these events (Németh 2016). While frameworks have been established to define the criteria to analyse the wide ranging legacy of events (Kassens-Noor 2016; Kassens-Noor et al. 2015), in terms of urban planning, research largely focuses on tangible physical urban impacts or the integration of mega-events into existing long-term strategies. However, such considerations for the governance and decision-making processes within mega-events should not be limited to the bidding or implementation phases, but also framed as a potential long-term legacy of the event.

While not all tools used to implement mega-events may be appropriate in the routine planning of cities during the legacy phase, certain aspects such as introducing widely shared strategic visions and plans as well as more inclusive participatory approaches could indeed be worthwhile practices to learn from and maintain. It is this issue of the institutional capacity to learn from the experience of hosting mega-events that this paper investigates. The inclusion of mega-events in strategic plans and long-term visions is by no means guaranteed (Deng, Poon, and Chan 2016) and is an aspect that cities must continue striving to better integrate (Jones 2019; Morandi and Di Vita 2017). As cities continue to focus on the long-term legacy of mega-events, this reflective learning can be an important aspect towards ensuring continuity and achieving the long-term goals attached to mega-events as part of legacy plans. This paper will investigate how one city, which hosted multiple mega-events, learned from each of these experiences and then examine what, if any, long-term learning took place to inform planning practice and decision making in the more than 15 years following the event.

The city of Genoa first hosted the 1992 Colombian Expo, followed by the 2001 G8 Summit and finally the 2004 European Capital of Culture (ECoC). The city's main impetus in hosting the events was to improve the historic city centre and reconnect to its waterfront in order to reposition the city as a cultural and tourist destination following several decades of decline and deindustrialization. The city used the events to focus heavily on its built heritage to promote a new image of the city based on its cultural heritage (Jones 2020). Beyond the vast physical transformation of the city, preparations for the event also introduced a number of new planning practices, namely developing a single, shared strategic vision that spanned political divides and included a wide range of local stakeholders. For Genoa, these represented the first attempts at utilizing public participation processes as well as taking a truly urban scale approach to city regeneration that went beyond the implementation of individual projects.

The following section of the paper shall briefly review some of the most relevant literature regarding the mega-event strategy, highlighting the hoped for benefits as well as the potential risks that bidding and host cities face. Section 3 then introduces Genoa and its three mega-events, describing the mechanisms introduced to plan and implement these events and the learning that was transferred from the 1992 Colombian Expo to the two later events. Section 4 critically examines the period following the event, analysing to what degree mega-event planning and governance came to inform practice over the last fifteen years since the close of the 2004 ECoC. Finally, the conclusions in Section 5 weigh the benefits of the mega-event strategy in the short term as well as review why the events failed to significantly impact the city's post-event planning and development, adding to calls in the literature for a more integrated mega-event planning approach by providing specific contributions from the case of Genoa.

2. The potential benefits and risks of the mega-event strategy

There are generally two common visions of hosting a mega-event: the optimistic tales of the Olympics of Los Angeles 1984 and Barcelona 1992 or the many cautionary tales such as the Athens 2004 and Sochi 2014 Olympic Games or the 2010 South Africa World Cup where newly created venues have become so-called 'white elephants' as they remain unused and empty following the event. Los Angeles has been cited as one of the few profitable games and the 'Barcelona Model' has become the template for cities hoping to use a mega-event to initiate an urban regeneration and rebranding programme (González 2011; Smith 2005, 2017). However, scholars emphasize that Barcelona's notable transformation resulted from a long-term planning process with the mega-event being just one contributing factor rather than the sole cause (Balibrea 2001; Degen and García 2012; Zimbalist 2015). The Liverpool 2008 ECoC has been framed as one of the most successful European Capitals of Cultures by both national and international experts and academics yet locals have been far more critical of the event (Sykes and Brown 2015). In reality, few events can be reductively classified as either a 'success' or 'failure'. A more useful differentiation would be to determine who benefits the most from these events, in which ways as well as who, on the other hand, is negatively affected. This review of the literature presents this range of potential positive and negative effects of mega-events as these issues significantly determine whether or not local decision makers choose to bid for these events as well as the challenges to then be addressed by event governance bodies.

A significant portion of literature on mega-events can be found within event and tourism studies, which often studies and aims to determine the degree of economic viability of these events over the long-term or their ability to attract tourism and promote city image internationally. In fact, long-standing definitions of mega-events often infer a set of positive attributes, such as those from Getz (Fayos-Solá 1998, 242) and Roche (2000) that broadly define events as having an extraordinary impact, mass appeal and increased visitor capacity. Of course, not all mega-events have in fact been successful in achieving such aims, but this narrative of success and transformation have become integral aspects of the way mega-events are understood and presented. Müller (2015), on the other hand, has devised a matrix to differentiate between the levels of events, from major to mega with giga events as the largest to better distinguish between events based on their actual delivery rather than depend on vaguely defined concepts.

Most scholars agree that the Los Angeles 1984 Olympics represent a turning point in the use and desirability of mega-events like the Olympics by demonstrating that the Games could in fact be profitable (Gold and Gold 2017). In the following decades, decision makers and special interests have promoted mega-events on the premise of growth and development, advertising the image of the city to a global audience and attracting mass tourism (Horne and Manzenreiter 2006; Roche 1994, 2002; Short 2008; Young and Wamsley 2005). Mega-event boosters or local supporters have also promoted events as a panacea with beneficial secondary effects spanning from physical improvements to the city through new mass infrastructural transport and mobility projects, new public facilities and broad socio-economic benefits (Preuss 2004). Though such rhetoric may be successful in building initial public support for mega-events, events are incapable of addressing each and every need of a city.

Though examples in the literature do highlight the wide range of effects that mega-events may have across different cities. One main area that mega-events intervene is in terms of service, transportation network or communication technologies that might otherwise be considered too complex or expensive but which are made possible due to the increased funding made available due to the event (Kassens-Noor 2013; Morandi and Di Vita 2017; Wilson 2011). Such efforts can also contribute to the goal of expanding local and regional economic diversification and growth (Gratton, Shibli, and Coleman 2006). One of the keys to the mega-event strategy, both in terms of financial terms as well as global visibility, is the level of tourism attracted both during and following events. Getz's definition, in particular, cites tourism as a key indicator of a mega-event and it also factors into Müller's (2015) calculation of event size. While the mega-event itself is framed as an important short-term attractor of tourism (Weed 2017), the anticipated long-term increase in tourism is often even more valued by cities (Faulkner 2003, 1993; Pyo, Cook, and Howell 1988). In fact, many of the pre-event economic assessments conducted include the expected increases in income from long-term tourism growth and are used to justify the high investments in mega-events (Archer 1982; Burgan and Mules 2000; Dwyer, Forsyth, and Spurr 2005; Rees 2000). Yet some studies have actually found that the presence of a mega-event reduced overall numbers of tourists as potential visitors stayed away to avoid large crowds and higher costs (Zimbalist 2015). In the case of the London 2012 Olympics, shops and vendors did not necessarily benefit from being located nearby to event venues as the high volume of spectators arrived to exclusively watch the event and then immediately left rather than spend time in those areas of the city, leading to economic losses for those local actors (Vlachos 2016).

Mega-events also have the potential to serve as a type of focusing event which unifies a range of actors and motivate them to cooperate when they might normally act in opposition to one another (McGillivray and McPherson 2012). However, there are many instances where mega-events have become quite politically divisive (Andranovich, Burbank, and Heying 2001). In recent years, both ends of the decision making spectrum have come together against mega-events to reject or cancel several bids to host the Olympic Games (Jones and Ponzini 2018). For example, a newly elected mayor of Rome and the city council of Budapest both cancelled their bids for the 2024 Games while public polls showed low support in Boston and Hamburg held a public referendum with a majority of citizens voting against hosting the Games. The long standing 'mega' qualities of mega-events have in many cases had negative impacts on the host city, largely due to the events going over already quite significant budgets and leaving abandoned venues and stadiums in their wake. Flyvbjerg and Stewart (2012) have found that every single Olympic Games since 1960 has gone over budget without exception. Zimbalist (2015) has also referred to Olympics and other mega-events as economic disasters, unable to recoup costs over the long-term.

While much of the literature focuses on sporting mega-events like the Olympics, there has been growing research into cultural mega-events as many small to medium scale cities have turned towards these events with the same intentions of promoting development, tourism and a new image of the city just as global cities have done with the Olympics or Expo (Jones 2020). Despite varying size and expenditure, the aspects of trans-nationality, exceptionality and regenerative abilities can be seen across differing types of mega-events, making them quite attractive to a wide range of cities. This can be seen as well in the recent spread and growth of City/Culture of Capital (CoC) events around the world (Green 2017). The success of the European Capital of Culture programme has in the past few decades has seen the creation of the Culture City of East Asia, Arab Capital of Culture, American Capital of Culture, Ibero-American Capital of Culture, UK City of Culture, Italian Capital of Culture and others. Due to this trend, more small and medium sized cities are turning to the mega-event strategy that has often been limited to larger globalized cities due to the physical as well as intangible regeneration promised.

It is in fact some of the very defining aspects of traditional mega-events, mainly their large expenditure and mass programme of development and infrastructure, that have led many cities to seek out the typically less expensive and more sustainable CoC events. To address such issues and respond to the decreasing interest in hosting the Olympic Games, the International Olympic Committee (IOC) created the Agenda 2020 that calls for proposals with reduced budgets and more sustainable proposals, a stark contrast from the long-standing dependence on ever bigger and more expensive Games (IOC 2014). The tenets of this programme can be seen in both the plans for the upcoming Paris 2024 and Milan 2026 Olympics that rely far more heavily on the existing infrastructure and venues of the city along with public spaces and even cultural heritage to host events and promote the city (Milano Cortina 2026, 2019; Paris 2024, 2018). This approach resembles, at last in terms of spatial distribution and use of existing venues, that which has long been used by the ECoC and other CoC events. It is worth noting this convergence of mega-event approaches diffusing activities throughout existing city fabric and which relies on prompting the city itself over newly icons created especially for the event. In this way, cities seem to be attempting to still acquire many of the desired benefits of promoting the city while avoiding the negatives of mass over-spending and unused venues that have plagued past host cities. This new trend of a diffused spread of events also gives greater urgency to studying cultural mega-events like the ECoC as cities have been implementing events in this manner for decades. In this way Genoa is an interesting case to study and which can contribute to understanding how city decision makers learn, or not, from the experience of bidding for, planning and implementing a mega-event.

3. Transferring knowledge between mega-events within the same city

Genoa turned to the mega-event strategy for reasons similar to most other cities. Local decision makers hoped to turn around decades of decline while improving the physical quality of the city and promoting a new heritage-based image of the city to a wider national and international audience. Following decades of degeneration and deindustrialization, the first steps of Genoa's urban renewal began with a few individual renewal projects realized in the late 1980s. Historically, Genoa was a stop on the Grand Tour, but during the twentieth century the city's port and industrial functions came to overshadow its wealth of heritage assets and the city's system of palaces and its related urban street system was essentially forgotten until Ennio Poleggi's (1968) research *The New Street: A Sixteenth Century Subdivision in Genoa.* In fact, Genoa has the largest preserved historic city centre in all of Europe and the Rolli System, historically, subdivided the palaces into categories to welcome varying levels of visitors to the city during the sixteenth and seventeenth centuries. The *strade nuove* or new streets were quite wide streets compared to the narrow medieval streets built to host the city's largest palaces of Genoa's noble families (Gorse 1997).

The first efforts to revitalize the degraded historic city centre focused on several key projects that included the University of Genoa Faculty of Architecture as well as the newly rebuilt Carlo Felice Theatre. However, these individual efforts did not have much of a broader impact and were unable to tackle the larger issues afflicting the city, namely the separation of the city centre from its port and waterfront area for over 80 years, both spatially as well as in terms of governance. In the 1980s, a new urban governance body was created, the Commissione Triporto, bringing together the municipality, the port authority and the regional government to jointly manage and agree on the historic port's future and how to introduce commercial and touristic functions (Bobbio 2005). This new shared governance body focused on the future of the city would lay the groundwork for the strategies the city would initiate with the Expo 1992 and build upon in the later 2001 G8 Summit and 2004 European Capital of Culture. These three large events were each used to concentrate on and improve this disconnect between the historic city centre and waterfront. The approach to regenerating the city would include a number of physical interventions in the city as well as recognizing and reaffirming the city's cultural heritage, particularly the series of Rolli Palaces (Jones 2017). These combined efforts were also intended to improve the overall quality and status of the city in order to attract new investments, development as well as tourists to the city through this heritage-led regeneration.

3.1. Experimenting with the mega-event strategy in the 1992 Expo

Compared to the earlier attempts at regeneration, the 1992 International Expo marked a significant difference in terms of scale and approach. Rather than relying on individual structures, the 1992 Expo was used to redevelop an entire area of the city: the waterfront. This began with the removal of the port walls that had previously separated the city from its waterfront throughout the twentieth century (Gastaldi 2016). The theme of the 1992 Expo also promoted the city's heritage by celebrating the 500th anniversary of Columbus discovering the Americas, one of Genoa's most famous residents. For the first time, the former port area was opened up to the city through a newly created public space that would not only host the activities of the expo, but remain as a long-term legacy for the city. The area was branded as the *Porto Antico* or Old Port and a number of previously abandoned and historic industrial buildings were restored and converted. The Magazzini del Cotone (Cotton Warehouse) is one such example and which continues to host a number of culture, leisure and commercial activities. The project represents one of the first efforts to embrace and re-use the city's industrial heritage, which had been largely overlooked due to the presence of the more prominent Rolli Palaces (Figure 1).

As an International Expo, the works completed in Genoa were smaller compared to those of a World Expo, but they still represented a key moment in the city's regeneration by completing the significant project of the newly created waterfront. Renzo Piano developed the plan for the Porto Antico that included a vast public space and, importantly, the new Genoa Aquarium. The model of establishing a new waterfront development with a large aquarium came directly from the example of Baltimore (De Marco 2006), one of the

Figure 1. Map of the regeneration works completed for the 1992 Expo in grey dash (source: Jones 2020).

most successful examples during the late 1980s and early 1990s of a waterfront regeneration (Falk 1986; Harvey 1991; Loftman and Nevin 1995). While a few restoration projects were also carried out across the city centre, the clear focus of the 1992 event was on the city's waterfront, which was the most critical area of the city in need of intervention at that time. However, local experts also criticized the planning of the 1992 Expo for not being strategic enough nor for considering the long-term legacy of the project in relation to the city centre (Gabrielli 2005). It is this lack of strategy and strong connections to the city centre in particular which would drive the city's approach in organizing the subsequent mega-events.

3.2. A strategic approach to planning the 2001 G8 and 2004 ECoC

Following the 1992 Expo, the city created a new governance body for the newly created Old Port area, the Società Porto Antico S.p.a., a public-private partnership with the municipality holding an 80% share (Gastaldi 2016). While the city attempted to shift to a new retail focus during the mid 1990s to help the city combat ongoing decline, these efforts would fail to bring about the desired improvements (Bobbio 2005). Ultimately, the city would once again turn to a mega-event to try to improve the city's circumstances and attempt to reverse continuing economic and social decline, though this time having learned several lessons from the experience of the 1992 Expo. Specifically, the planning for the 2004 ECoC sought to be more strategic by involving a wide range of local stakeholders, developing a larger vision for the city and diffusing the works associated with the event throughout the city centre, and beyond, rather than concentrating in a single zone of the city. This approach was supported by Pasqual Maragall, the former Mayor of Barcelona known for leading the successful 1992 Bilbao Olympic Games and advised Genoa to organize a strategic conference to develop a shared vision for the event with key stakeholders. The strategic conference was organized shortly after Genoa was awarded the 2004 ECoC and became the key moment to plan the future development of the city and to utilize the 2004 ECoC as the mechanism to achieve many city goals.

The *1999 Strategic Conference* and subsequent *2000 Operation Plan for the Historic Centre* (Comune di Genova 1999, 2000) intended to apply the lessons learned from the experience of hosting the 1992 Expo by developing a strategic approach that would ensure the actions taken for the event responded to the needs and goals of the city. The Strategic Conference brought together a diverse range of actors from differing sectors, institutions, levels and even political affiliation to develop a singular vision for Genoa. The eventual managing body created to guide the event and associated projects, the Genova 2004 S.r.l., reflected this joint venture with representatives from the municipality, province, region, chamber of commerce, port and universities. Interviews with event organizers revealed that the ECoC also served as moment for the general public to participate and propose project ideas, a new tactic for the city. Through these various approaches, Genoa embedded the planning of the mega-event within a set of wider city plans and strategies from the early planning phases that brought together actors to work towards a common shared goals of recovering the historic city centre to unify the polycentric city of Genoa while also promoting it as a cultural destination.

The *2000 Operative Plan for the Historic City Centre* generated from the *1999 Strategic Conference* laid out the specific projects to be completed for the 2004 ECoC. Due to the integrated nature of the plans generated, the city incorporated these plans with other ongoing urban plans and funding schemes to be able to complete as many works throughout the city centre as possible. Genoa would use funds from the local, regional, national and even European levels by coordinating works for the 2004 ECoC through the CdQ, PRU, CIV, PRUSST and EU Urban II funds (see Table 1). The city would also take advantage of the 2001 G8 Summit to secure increased state funding and complete additional regeneration and restoration projects. The Operative Plan originally proposed 31 specific projects to be finished by 2001–2002 and another 70 to be prepared for the opening of the ECoC in 2004. Though the Strategic Conference envisioned plans only for the 2004 ECoC, as the city had not yet been awarded the 2001 G8, restorations on some of the most important palaces along with the pedestrianization of a number of key throughways throughout the city centre were expedited and completed early for the 2001 deadline. Overall, the *Operative Plan for the Historic City Centre* covered nearly the entire 113 ha of the city centre, focusing on much of the city's urban heritage, restoring both historic structures and public spaces in tandem (see Figure 2). To accomplish such a high number of works within such a short period, the Operative Plan interconnected projects not only between the two events but also with the variety of existing plans and funding sources from regional, national and EU levels (Comune di Genova 2000). In this way, the 2001 G8 and 2004 ECoC served as a focusing event that brought together a range of actors and funding schemes to work towards a single coordinated goal. This tightly unified approach also encouraged private actors to fund a number of restoration works on historic structures in the city centre thereby contributing to the overall impact.

The city worked to recover the city centre through a mass urban regeneration programme spanning nearly the entire historic city centre and supported its cultural offerings by establishing a new museum system including a series of museum poles

Table 1. The main city plans and funding initiatives in Genoa prior, during and following the 2004 ECoC (source: elaboration of Jones 2020).

Years	Urban plan/programme/event
1992	Expo 1992
1994–1999	EU Urban I, European Commission and ERDF
1997–2007	CdQ – Contract of the Neighbourhoods, co-national/regionally funded
1998–2008	PRU – Urban Redevelopment Program, publicly and privately funded
1999	Strategic Conference
2000	Operative Plan for the Historic City Centre
2000	PUC – Urban Plan of the City
2000–2006	CIV – Integrated Centre Streets
2000–2006	EU Urban II
2001	G8 Summit
2002–2012	PRUSST – Urban Redevelopment Program and Sustainable Development of the Territory, promoted by the Ministry of Public Works
2004	European Capital of Culture
2006	UNESCO World Heritage Listing
2007–2013	ROP – Regional Operational Program, funded by the EU and ERDF
2008–2011	TSN 1 & 2 – Territory Projects in the Joint Territories
2014–2020	ROP – Regional Operational Program, funded by the EU and ERDF
2015	PUC – Urban Plan of the City
2017	Strategic Plan of the Metropolitan City

Figure 2. Map of the urban regeneration works completed for the 2001 G8 and 2004 ECoC in grey dash (source: Jones 2020).

that brought together existing institutions along with several newly created museums (Jones 2020). It is important to note that at the time of planning the 2004 ECoC, Genoa was not known as either a city of culture or tourism. Thus the heavy promotion of the city as a cultural destination represented a new idea and image of the city. In many ways, the intention was to continue and essentially complete the works begun on the city's waterfront for the 1992 Expo by improving connections to the city centre to create a more unified hub that could serve as a true core for the polycentric distribution of Genoa. Rather than relying primarily on the construction of new infrastructure or iconic venues, the city focused on revitalizing existing public spaces and inserting new cultural centres/museums within this existing fabric. In the city centre, plans focused along key urban corridors to improve accessibility throughout the city centre and waterfront. These included the new streets of Via Balbi and Via Garibaldi which host the grandest of the Rolli Palaces and the heritage assets of the city were included as key resources within this urban revitalization, to both visitors and citizens alike back into and through the city centre.

As Figure 2 shows, this strategy took a far more comprehensive approach to the city centre and waterfront compared to the works completed for the 1992 Expo which concentrated almost exclusively on the Porto Antico (see Figure 1). In this way, the plans for the 2001 G8 and 2004 ECoC built on the existing works already completed a decade earlier while also learning key lessons from the shortcomings of the previous approach. In preparation for the 2001 G8, over 100 restoration projects and several key pedestrian pathways were completed. The 2004 ECoC would see an additional 60 heritage-based projects completed including façade restorations, street improvements and the renewal of public spaces (Jones 2017, 2020; Carbonara 2001; Farini, Mantero, and Salvarani 2004; Pittarello 2001). Additionally, for the 2004 event, the city also created several new cultural infrastructures along with new uses to several of the restored spaces. These included the new Galata Museum of the Sea, the Biosphere by Renzo Piano, a botanical garden located on the waterfront and connected to the aquarium, as well as the new Palazzo Verde, a former industrial warehouse that was converted into an education and activity centre focused on sustainability issues.

The 2004 ECoC also established a new museum system that connected many existing museums as well as incorporated several of the palaces that had been restored for the 2001 G8. Through its 12-year mega-event strategy, the city of Genoa well utilized many of the potential positive aspects of the mega-events, namely the increased funding, visibility and unification of actors to integrate the projects within a clear planning approach. While two of these mega-events, the 2001 G8 and 2004 ECoC were essentially executed in tandem, the planning of these events learned significantly from the approach of the 1992 Expo. This approach helped the city to avoid many of the pitfalls associated with mega-events, particularly overspending or large iconic structures left unused following the close of the event as all of the newly created cultural centres and museums have remained active. Despite these many positive outcomes, the city was at the same time not entirely successful in reversing trends of economic decline and depopulation as hoped for (Città Metropolitana di Genova 2017). While it seems clear that there was a transfer of knowledge when it came to hosting mega-events, the next section will look into the challenges of maintaining such a strategic and integrated approach in the absence of a mega-event.

4. A failure to learn from the mega-event strategy over the long term

There are several legacies that emerge from Genoa's use of the mega-event strategy. First, the city created and implemented a shared strategic vision centred on a comprehensive urban regeneration programme, first through the rehabilitation of the waterfront area of the old port and then through reinforcing connections to the city centre and embracing the city's urban heritage. This approach led to developing a previously non-existent tourism economy in the historic city centre, as can be seen through consistent growth in the numbers of overnight stays in the city at a rate higher than the national average (Jones 2020). Due also to the strong focus on heritage restoration, a significant portion of the city was also recognized as a UNESCO World Heritage Site in 2006 (UNESCO 2006). Additionally, the newly created City Museum system has become of the main city attractors, consistently growing in numbers of total visitors since 2005 (Città Metropolitana di Genova 2017). The city accomplished many, though not all, of the goals it set in hosting this series of mega-events by learning to embed them in long-term strategic processes developed through inclusive processes involving city stakeholders as well as public participation. As with many mega-events, the ECoC was framed as the end point to these efforts with the city then addressing other areas of need in subsequent years.

As happens all too often following most mega-events, the event managing body, 2004 Genova S.r.l. was disbanded with no legacy organization left to continue managing cultural events or focus on the city centre. Interviews with the former director of the agency revealed the lack of any legacy plan or organization that might have continued the agency's efforts. This ultimately resulted in the loss of institutional knowledge and capacity building that had taken place over the seven years of planning and preparing for the events. Interviews with planning officers from the local municipality collaborated on projects carried out for the events confirmed the loss of expert knowledge as these internal networks of cooperation were lost in the period following the event. These changes in part took place due to the arrival of a new mayor in 2007 that shifted focus from the city centre to other areas of the city. After so many years addressing the issues in the centre, it was natural for Genoa to turn its attention to other areas. However, when one looks to the planning approach as well as the types of projects proposed in the years following the mega-events one observes a clear shift in tactic that lacks a shared strategic vision, public participation and an urban approach that carefully integrates into existing urban fabric, the keys to the strategy developed for the 2004 ECoC.

As Section 3 demonstrated, the approach to hosting the ECoC derived from the *1999 Strategic Conference* and *2000 Operative Plan for the Historic City Centre* that brought together many other funding sources to augment the works completed. Even without the additional funding and emphasis, this strategic and inclusive approach could have become a lesson learned through the experience of hosting the mega-events for the city to continue implementing. However, by examining the planning practice of the city we can describe three distinct periods, as illustrated in Figure 3. The initial 'experimental' phase mainly consisted of the first attempt at using a mega-event to transform the city, implemented largely as a single and contained programme separate from other plans or projects. The period from 1997 until 2004 represents a much more active period in terms of overlapping plans and projects which were largely unified in their aims and often collaborated to complete projects. Following that phase, the Regional Operational Programs (ROP), funded by the EU and ERDF, as well as the Territory Projects in the Joint Territories TSN 1 & 2 show the city's shifting focusing to regional and territorial concerns, but there is not the same concentration or integration of plans as found during the previous period. Additionally, there was no larger shared vision to unite different actors and stakeholders or to unify these regionally focused efforts. While the *1999 Strategic Conference* was fundamental in developing the plans for the 2004 ECoC, the city would not publish its next strategic plan until 2017, nearly twenty years later. While this later plan was developed through participative processes across the metropolitan area, it was clearly not a practice learned from the much earlier mega-event but a newly reintroduced practice.

In the last twenty years, several large-scale development projects have also been proposed to encourage Genoa's growth and development. Upon closer inspection, the

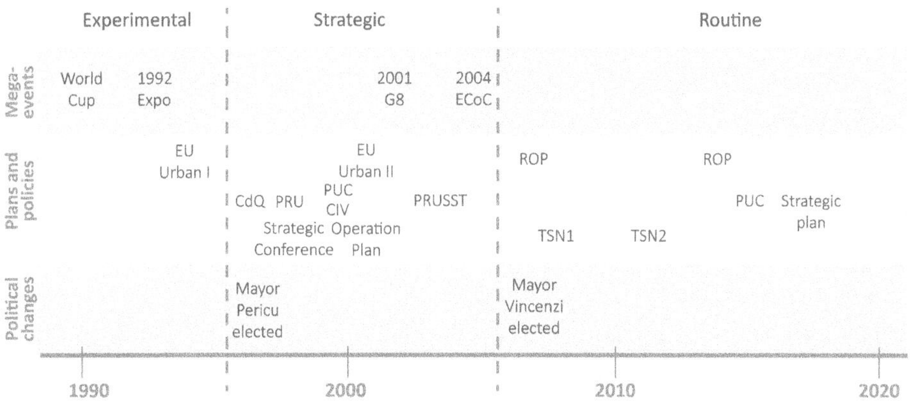

Figure 3. Three phases of planning practice during and following Genoa's mega-event strategy (source: author).

proposed plans take a different approach to urban regeneration than that implemented through the mega-events. Three highly visible projects have been the Ponte Parodi, a multi-use entertainment and shopping centre originally intended as an ECoC project, the Erzelli GREAT Campus, a technology district located in the upper hills in the Sestri neighbourhood, and Blueprint, a reorganization of the existing fairgrounds and port area on the eastern edge of the waterfront by Renzo Piano (see Figure 4). Of these three, only the Erzelli GREAT Campus has begun construction with some office buildings now open. While both Ponte Parodi and Blueprint continue to deal with the city's waterfront and port activities, they are much more inward looking and stand-alone projects that lack significant connections to the existing city fabric surrounding them. While the Erzelli GREAT Campus contributes towards the city's current regional and territorial focus of supporting other areas beyond the city centre, it also fails to engage with its surrounding area. Its urban design reminiscent of most BCDs developed in the late 1980s and 1990s, a strategy which Genoa also utilized with its SOM designed San Benigno Tower nicknamed *Il Matitone* or the Big Pencil from 1992. While it is not yet possible to know the eventual impact of these projects, their planning and implementation follows the usual conventions of privately driven large development projects rather than the strategically planned and urban approach utilized for the 2001 G8 and 2004 ECoC. While the Erzelli GREAT Campus and Blueprint are referenced in the 2017 Metropolitan City Strategic Plan, they existed as independent plans years earlier and were not originally developed as part of a wider approach focused on a shared vision. In comparing the differences of planning approach and specific details of these plans, it seems clear that Genoa's experience of using the mega-event strategy over a 12-year time span did not leave a lasting impact on post-event planning processes.

One of the keys of success to Genoa's mega-event strategy was utilizing the event as a focusing event that both organized and motivated a diverse range of public and private actors, an impressive feat that many cities do not achieve. This aspect along with Genoa's strategic oriented approach would have been valuable planning practices for the city to learn from and continue implementing beyond the close of the 2004

Figure 4. The spatial layout of the three large-scale projects (in grey dash): Erzelli GREAT Campus, Ponte Parodi and Blueprint (source: author).

ECoC. Figure 3 divides the city's planning practice into three phases based on the analysis of city plans and approach urban regeneration and development. The 1992 Expo was a largely isolated phenomenon, unconnected to a wider vision or strategy for the city. It served as a kind of new experiment for the city, testing the possibilities of this approach by focusing on a key area of the city, the waterfront. Having learned several key lessons, decision makers strategically planned and implemented the 2001 and 2004 events by bringing together a number of separate plans, initiatives and funding schemes, focusing all of them on the goals established through the strategic conference. While this strategy was successful in many regards, one of its main shortcomings was in framing the 2004 ECoC as the end of these regeneration process, rather than as a key step within an even longer-term vision for the city. Following the mega-events the city then seems to have returned to the typical 'routine' of planning and development, more reminiscent of past projects implemented in the 1990s than the strategic and more urban approach used in planning the 2001 G8 and 2004 ECoC.

5. Conclusions: calls for a more integrated and reflective mega-event strategy

It is clear that while the series of mega-events Genoa hosted did bring about significant change to the city, both physically and in its cultural and tourism offerings, it is difficult to find any long-term impact or learning that took place in terms of planning approach. However, Genoa was successful in learning from the planning and organization of the 1992 Expo and adjusted accordingly its approach to the later events. When faced with the similar situation of the mega-event, which provides strict deadlines as well as additional funding possibilities, the city managed to work in a more strategic way through a shared vision. However, in the absence of an event, the city seems to have returned to more standardized large project development processes that lacked the more inclusive approach of planning that occurred for the 2004 ECoC. While in the short-term the city clearly benefited in a number of ways through the mega-event strategy, its effects on planning practice are quite limited from a long-term perspective. Many of the outputs of the event, particularly regarding the development of the city's cultural offerings have been quite successful, becoming an important part of the city's economy in terms of tourism attraction, yet the innovative planning approaches that helped to generate these changes were not continued.

As often occurs, was no opportunity for many of the actors involved in these process to contribute their learning to future planning initiatives as the event management bodies were dissolved or municipality networks were broken up, leading to a loss of institutional capacity. In this way, Genoa seems to have returned to normal routine in terms of planning with the exceptional routines or practices used to organize the mega-event deemed no longer necessary going forward. While cities cannot continuously rely upon special legislation or extreme funding to work around regular procedure, there may be worthwhile aspects to be maintained. In the case of Genoa, the city could have continued working to implement a strong collaborative approach embedded within a comprehensive strategic vision. While the event did provide necessary additional funding normally not available, part of the budget came from other pre-existing public funds that were then implemented in a strategic and coordinated way, an approach that could have continued to benefit the city. Such an integrated planning approach that brings together different projects and funding schemes could have also helped in addressing the regional and territorial issues the city turned towards following the event.

In reflecting on the case of Genoa, there are several valuable takeaways that other cities could work to implement when pursuing the mega-event strategy. While most scholars would agree that the plans for the event benefit from being included within a wider strategic vision, cities should be careful to avoid framing the event as the final step in such processes. Instead, strategic visions and plans should also frame the legacy of the event not as a separate phase but an ongoing continuation of the plan. This reframing of mega-events as an important step rather than the final one can also temper expectations for the mega-event and not risk over-promising more than it can possibly deliver. Framing the mega-event planning process in this way could also help in maintaining the various networks and governance structures involved in organizing the event rather than losing their collective knowledge and experience at the close of the event. Even without the hard deadlines and additional funding of the mega-event, this repositioning of the mega-event within long-term processes can help to maintain a strategic and forward thinking approach to planning rather than just a return to routine practice.

A second key aspect is for cities to recognize and value the institutional knowledge and experience gained by the new governance structures that plan and manage the event. As seen in Genoa, these might be in the form of new agencies or foundations as well as more informal networks between municipal departments and actors that traditionally do not collaborate together. The formation and subsequent loss of such networks is unfortunately quite common amongst ECoCs in particular, where local actors tend to be far more involved compared to other mega-events like the Olympics that utilize external experts. These governance bodies can be considered as part of the legacy of the event and could be utilized to address and plan for other issues facing the city. While some cities have begun funding ongoing cultural offices following the events, as seen in Liverpool or Wrocław, that continue organizing cultural activities, rarely does this take place more broadly in terms of planning practice. One way this could be achieved would be to have municipality staff embedded within governance structures, first to foster greater cohesion between the event management and city planning, but also to ensure that learning and experience are retained as those staff eventually return to work for the municipality. While changing political leadership may inevitably alter city plans or strategies, a valuing of the mega-event planning process itself and taking steps to learn from it could help ensure that such approaches are retained regardless of the eventual focus of the city.

While seen only in a limited capacity in the case of Genoa, most ECoCs since have increasingly turned to public participation as a key component in the bidding phase of hosting the event. However, while the event can often serve as the impetus to turn towards more inclusive processes, they are not necessarily carried throughout the implementation and legacy phases, as witnessed in Genoa. In cities where there already exists a tradition of public participation prior to the event, they can instead serve as an opportunity to expand these efforts or include communities typically left out. While events like the Olympics have yet to embrace public participation in the bidding phase, it has become a key component for cities bidding to host the ECoC that demonstrates not only support for the bid but also that the bid reflects the community as a whole. Yet such processes should not be viewed solely as a technique to win events, but strive to learn from these experiences and include them within typical planning processes.

In using the mega-event strategy, the case of Genoa shows the risks of cities becoming overly reliant on the exceptional routines that mega-events can provide rather than learning from the process and applying those lessons to regular planning practice. Genoa hosted these mega-events in order to physically regenerate different areas of the city and to promote a new image, an approach shared by many other cities hosting mega-events and which could learn very similar lessons. Even in cities that intervene very little on their built environments through such events can surely also learn important lessons from the experience of planning mega-events and reframe both how they come to view mega-events within long-term plans as well intentionally valuing and learning from the mega-event planning process itself. In the ongoing discussions of legacy, both within the literature and by cities themselves, more consideration should be given for the planning capacity that can be increased throughout the experience, reflecting upon it in order to learn and inform future planning procedure. By focusing solely on the physical or socio-economic legacy of mega-events, cities risk losing key knowledge built up over years and not easily replicated.

Disclosure statement

No potential conflict of interest was reported by the author(s).

ORCID

Zachary Mark Jones http://orcid.org/0000-0001-6423-957X

References

Andranovich, G., M. Burbank, and C. Heying. 2001. "Olympic Cities: Lessons Learned from Mega-Event Politics." *Journal of Urban Affairs* 23 (2): 113–131. doi:10.1111/0735-2166.00079.

Archer, B. 1982. "The Value of Multipliers and Their Policy Implications." *Tourism Management* 3 (4): 236–241. doi:10.1016/0261-5177(82)90044-9.

Balibrea, M. 2001. "Urbanism, Culture and the Post-industrial City: Challenging the "Barcelona Model."." *Journal of Spanish Cultural Studies* 2 (2): 187–210. doi:10.1080/14636200120085174.

Basso, M. 2015. *Governare "routines eccezionali" I grandi eventi e la città contemporanea, tra eccezione e regola: Forme, pratiche ed effetti sul policy-making. Considerazioni attorno ai casi di Londra.* Milano: Università IUAV di Venezia.

Basso, M. 2017. *Grandi eventi e politiche urbane: Governare "routine eccezionali" un confronto internazionale.* Milan: Guerni e Associati.

Bobbio, R. 2005. "Complessità di rapporti e iniziative di integrazione fra la città e il porto di Genova." *Portus* 10: 35–41.

Burgan, B., and T. Mules. 2000. "Event Analysis: Understanding the Divide Between Cost Benefit and Economic Impact Assessment." In *Events Beyond 2000: Setting the Agenda*, edited by J. Allen, R. Harris, L. K. Jago, and A. Veal, 46–51. Sydney: Australian Centre for Event Management.

Carbonara, G. 2001. "Un bilancio positivo." *Arkos* 2 (1): 6–10.

Città Metropolitana di Genova. 2017. *Piano strategico della città metropolitana.* https://pianostrategico.cittametropolitana.genova.it/sites/default/files/psm2017/All.1_PSM%20aprile%202017.pdf (on 12 September, 2019)

Comune di Genova. 1999. 1 Sessione "Genova: Città della cultura e del turismo." Genova: Le Vie Del Mediterraneo All'europa. Conferenza strategica della città di Genova, Genoa.

Comune di Genova. 2000. *Il PIano Operativo per il Centro Storico*, Genova.

De Marco, L. 2006. Which Image(s) for Genoa as a World Heritage Site?" 9th US/ICOMOS International Symposium "From World Heritage to Your Heritage," Newport.

Degen, M., and M. García. 2012. "The Transformation of the 'Barcelona Model': An Analysis of Culture, Urban Regeneration and Governance." *International Journal of Urban and Regional Research* 36 (5): 1022–1038. doi:10.1111/j.1468-2427.2012.01152.x.

Deng, Y., S. Poon, and E. Chan. 2016. "Planning Mega-Event Built Legacies–A Case of Expo 2010." *Habitat International* 53: 163–177. doi:10.1016/j.habitatint.2015.11.034.

Duan, J., X. Yi, and K. Kunzmann. 2016. "Big Events and Urban Governance in China: Experience from the Youth Olympic Games in Nanjing." *disP – The Planning Review* 52 (1): 18–30. doi:10.1080/02513625.2016.1170486.

Dwyer, L., P. Forsyth, and R. Spurr. 2005. "Estimating the Impacts of Special Events on an Economy." *Journal of Travel Research* 43 (4): 351–359. doi:10.1177/0047287505274648.

Falk, N. 1986. "Baltimore and Lowell: Two American Approaches." *Built Environment* 12 (3): 145–152.

Farini, A., A. Mantero, and F. Salvarani. 2004. "Il recupero dell'antico waterfront." *Arkos*, 12–13.
Faulkner, B. 2003. *Progressing Tourism Research*. Edited by L. Fredline, L. Jago, and C. Cooper. Clevedon, UK: Channel View Publications.
Faulkner, H. W. 1993. *Evaluating the Tourism Impacts of Hallmark Events*. Canberra: Bureau of Tourism Research.
Fayos-Solá, E. 1998. "The Impact of Mega Events." *Annals of Tourism Research* 25 (1): 241–245. doi:10.1016/S0160-7383(97)00083-2.
Flyvbjerg, B., and A. Stewart. 2012. "Olympic Proportions: Cost and Cost Overrun at the Olympics 1960-2012." *Saïd Business School Working Papers*, 1–23.
Gabrielli, B. 2005. "Genova, un piano strategico di natura operativa ei suoi esiti." *Urbanistica* 126: 57–63.
Gastaldi, F. 2016. "Genoa, a Success Story!." In *Waterfronts Revisited: European Ports in a Historic and Global Perspective*, edited by H. Porfyriou and M. Sepe, 123–133. New York: Routledge.
Gold, J., and M. Gold. 2017. "The Enduring Enterprise: The Summer Olympics, 1986–2012." In *Olympic Cities: City Agendas, Planning, and the World's Games, 1896–2020*, edited by J. R. Gold and M. M. Gold, 21–63. New York: Routledge.
Gold, M. 2017. "Athens 2004." In *Olympic Cities: City Agendas, Planning, and the World's Games, 1896–2020*, edited by J. Gold and M. Gold, 333–358. New York: Routledge.
González, S. 2011. "Bilbao and Barcelona 'in Motion'. How Urban Regeneration 'Models' Travel and Mutate in the Global Flows of Policy Tourism." *Urban Studies* 48 (7): 1397–1418. doi:10.1177/0042098010374510.
Gorse, G. 1997. "A Classical Stage for the Old Nobility: The Strada Nuova and Sixteenth-Century Genoa." *The Art Bulletin* 79 (2): 301–326. doi:10.1080/00043079.1997.10786737.
Gratton, C., S. Shibli, and R. Coleman. 2006. "The Economic Impact of Major Sports Events: A Review of Ten Events in the UK." *The Sociological Review* 54 (2_suppl): 41–58. doi:10.1111/j.1467-954X.2006.00652.x.
Green, S. 2017. *Capitals of Culture: An Introductory Survey of a Worldwide Activity*. Prasino. http://prasino.eu/wp-content/uploads/2017/10/Capitals-of-Culture-An-introductory-survey-Steve-Green-October-2017.pdf
Harvey, D. 1991. "The Urban Face of Capitalism." In: *Our Changing Cities*, edited by J. F. Hunt, 50–66, Baltimore, NJ: Johns Hopkins University Press.
Horne, J., and W. Manzenreiter. 2006. "An Introduction to the Sociology of Sports Mega-events1." *The Sociological Review* 54 (s2): 1–24. doi:10.1111/j.1467-954X.2006.00650.x.
IOC. 2014. *Olympic Agenda 2020: 20 + 20 Recommendations*. Lausanne: International Olympic Committee.
Jones, Z. 2017. "Synergies and Frictions Between Mega-Events and Local Urban Heritage." *Built Heritage* 1 (4): 22–36. doi:10.1186/BF03545655.
Jones, Z. 2019. "Policy and Practice Integrating Cultural Events and City Agendas: Examples from Italian/UK Practice." *Town Planning Review* 90 (6): 587–599. doi:10.3828/tpr.2019.38.
Jones, Z. 2020. *Cultural Mega-Events: Opportunities and Risks for Heritage Cities*. Abingdon: Routledge.
Jones, Z., and D. Ponzini. 2018. "Mega-events and the Preservation of Urban Heritage: Literature Gaps, Potential Overlaps and a Call for Further Research." *Journal of Planning Literature* 33 (4): 433–450. doi:10.1177/0885412218779603.
Kassens-Noor, E. 2013. "Transport Legacy of the Olympic Games, 1992–2012." *Journal of Urban Affairs* 35 (4): 393–416. doi:10.1111/j.1467-9906.2012.00626.x.
Kassens-Noor, E. 2016. "From Ephemeral Planning to Permanent Urbanism: An Urban Planning Theory of Mega-Events." *Urban Planning* 1 (1): 41–54. doi:10.17645/up.v1i1.532.
Kassens-Noor, E., M. Wilson, S. Müller, B. Maharaj, and L. Huntoon. 2015. "Towards a Mega-Event Legacy Framework." *Leisure Studies* 34 (6): 665–671. doi:10.1080/02614367.2015.1035316.
Lauermann, J. 2019. "The Urban Politics of Mega-Events." *Environment and Society* 10 (1): 48–62. doi:10.3167/ares.2019.100104.

Loftman, P., and B. Nevin. 1995. "Prestige Projects and Urban Regeneration in the 1980s and 1990s: A Review of Benefits and Limitations." *Planning Practice & Research* 10 (3–4): 299–316. doi:10.1080/02697459509696280.

McGillivray, D., and G. McPherson. 2012. "Mega Events: Neoliberalized Vehicle or Opportunity for Strategic Global Leadership." *Journal of Leadership, Accountability and Ethics* 9 (5): 80–91.

Milano Cortina 2026. 2019. *Milano Cortina 2026 Candidature Dossier*. https://www.milanocortina2026.coni.it/en/files/dossier/28-dossier-english-version.html

Morandi, C., and S. Di Vita. 2017. *Mega-Events and Legacies in Post-Metropolitan Spaces: Expos and Urban Agendas*. Cham: Palgrave MacMillan.

Müller, M. 2014. "After Sochi 2014: Costs and Impacts of Russia's Olympic Games." *Eurasian Geography and Economics* 55 (6): 628–655. doi:10.1080/15387216.2015.1040432.

Müller, M. 2015. "What Makes an Event a Mega-Event? Definitions and Sizes." *Leisure Studies* 34 (6): 627–642. doi:10.1080/02614367.2014.993333.

Németh, Á. 2016. "European Capitals of Culture–Digging Deeper into the Governance of the Mega-Event." *Territory, Politics, Governance* 4 (1): 52–74. doi:10.1080/21622671.2014.992804

Paris 2024. 2018. *Paris 2024 Presents an Optimised Olympic Games Venue Concept*. https://www.paris2024.org/app/uploads/2018/10/press_release_paris_2024_-_update_paris_2024_presents_an_optimised_olymic_games_venue_concept.pdf

Pittarello, L. 2001. "Molti lavori per un complessivo intervento di restauro urbano." *Arkos* 2 (1): 11–17.

Poleggi, E. 1968. *Strada nuova: Una lottizzazione del Cinquecento a Genova*. Genova: Sagep.

Preuss, H. 2004. *The Economics of Staging the Olympics: A Comparison of the Games, 1972–2008*. Cheltenham: E. Elgar.

Pyo, S., R. Cook, and R. Howell. 1988. "Summer Olympic Tourist Market—Learning from the Past." *Tourism Management* 9 (2): 137–144. doi:10.1016/0261-5177(88)90023-4.

Rees, M. 2000. Issues in Evaluation: EventsCorp's Perspective. In *Events Beyond 2000: Setting the Agenda*, 75, Sydney.

Roche, M. 1994. "Mega-events and Urban Policy." *Annals of Tourism Research* 21 (1): 1–19. doi:10.1016/0160-7383(94)90002-7.

Roche, M. 2000. *Mega-events and Modernity: The Olympics, Internationalism and Supernationalism: International Sports Events and Movements in the Inter-war Period*. New York: Routledge.

Roche, M. 2002. *Megaevents and Modernity: Olympics and Expos in the Growth of Global Culture*. London: Routledge.

Short, J. 2008. "Globalization, Cities and the Summer Olympics." *City* 12 (3): 321–340. doi:10.1080/13604810802478888.

Silvestre, G. 2009. "The Social Impacts of Mega-Events: Towards a Framework." *Esporte e Sociedade* 4 (10): 1–26.

Smith, A. 2005. "Conceptualizing City Image Change: The 're-Imaging' of Barcelona." *Tourism Geographies* 7 (4): 398–423. doi:10.1080/14616680500291188.

Smith, A. 2017. "Urban Regeneration." In *Olympic Cities: City Agendas, Planning, and the World's Games, 1896–2020*, edited by J. Gold and M. Gold, 3rd ed., 217–229. London: Routledge.

Sykes, O., and J. Brown. 2015. "European Capitals of Culture and Urban Regeneration: An Urban Planning Perspective from Liverpool." *Urbanistica* 155: 79–85.

UNESCO. 2006. *Decisions Adopted at the 30th Session of the World Heritage Committee*. United Nations Educational, Scientific and Cultural Organization. http://whc.unesco.org/archive/2006/.

Vlachos, P. 2016. "The 2012 Olympics and Small Local Business: A 5-Year Longitudinal Study of South-East London." In *The London Olympics and Urban Development*, edited by G. Poynter, V. Viehoff, and Y. Li, 176–192. Abingdon: Routledge.

Weed, M. 2017. "Olympic Tourism." In *Olympic Cities: City Agendas, Planning, and the World's Games, 1896–2020*, edited by J. Gold and M. Gold, 3rd ed., 230–252. London: Routledge.

Wilson, M. 2011. "Event Engineering: Urban Planning for Olympics and World's Fairs." In *Engineering Earth*, edited by S. D. Brunn, 1045–1055. Dordrecht: Springer Netherlands. doi:10.1007/978-90-481-9920-4_59

Wu, Y., X. Li, and G. C. Lin. 2016. "Reproducing the City of the Spectacle: Mega-Events, Local Debts, and Infrastructure-led Urbanization in China." *Cities* 53: 51–60. doi:10.1016/j.cities.2016.01.004.

Young, Kevin, and Kevin Wamsley, eds. 2005. *Global Olympics: Historical and Sociological Studies of the Modern Games*. Amsterdam: Elsevier.

Zimbalist, A. 2015. *Circus Maximus: The Economic Gamble Behind Hosting the Olympics and the World Cup*. Washington DC: Brookings Institution Press.

A mega-event in a small city: community participation, heritage and scale in the case of Pafos 2017 European Capital of Culture

Evanthia Dova, Angeliki Sivitanidou, Natia R. Anastasi and
Julia Georgi-Nerantzia Tzortzi

ABSTRACT
Pafos's 'Open Air Factory' strategy for the 2017 ECoC was a simple yet imaginative concept born out of a strong desire to re-draw the city's small-town character, diversify its economy, and steer social and cultural development along new paths. In order to counter the city's limited economic resources and cultural infrastructure, the event's vision was based on the two assets that Pafos has in abundance: a rich cultural heritage and a community eager for change. This paper tracks the uneven path of community involvement during the ECoC bidding, planning, implementation and post-implementation stages, and looks at urban heritage as the shared field of convergence for the multiple modes and degrees of participation in the realization of a mega-event. The city's small size allows for original insights into the opportunities and limitations of the intersections between community, heritage and scale in the case of mega-event-based urban regeneration.

Introduction

This paper focuses on the degree of citizen involvement in cultural events in the Pafos 2017 ECoC and its fluctuations from the bidding to the post-event stages. With the agenda of participation gaining prominence in the ECoC scheme, we investigate how Pafos's size facilitated and hindered the various modes of community engagement as the mega-event evolved.

As a small city[1], Pafos has a scale that allows easier access to local administration and less mediated participation in public affairs. Pafos, '… a small city in a small province in one of the smallest countries of the European Union' (Grizzo et al. 2011, 18), used its scale as the defining factor in planning, funding and delivering the 2017 ECoC mega-event. Pafos took advantage of its size to present itself as a paradigm where the impact of the mega-event would be strong enough to introduce new directions for its stymied social and cultural development. Local involvement provided a dynamic start to the ECoC effort and succeeded in uniting the Pafos residents around a common cause, in creating a sense of joint ownership for the project and in providing in-depth knowledge of the region's strong assets, problematic areas, and potential development directions (Dova 2013).

Cultural heritage provided the space – both literally and metaphorically- on which the various local groups came together. The spatial and the cultural components of Pafos's 'Open Air Factory' bid were shaped by open volunteer workshops, which concentrated on the Pafos region spatial heritage, and its long cultural traditions and history, using them as springboards for further urban projects and cultural productions. The main idea behind Pafos's cultural 'factory' was its *openness*, a catchword which alluded to both open-air locations and unlimited access to cultural participation and co-production. The bidding effort also managed to establish a culture of organized volunteering, which, for Pafos, was without precedent, and which strengthened the bid's credibility. During the planning phase, public involvement was noticeably limited; however, it was reanimated in the year of the event, when locals found their way to the ECoC projects through organizing a few events, but mostly as audiences and volunteers. The energized participation of the locals in the ECoC events has helped characterize the event as successful, in terms of enhancing the region's cultural infrastructure, capacity and audience development. Yet the lack of an operational legacy plan -despite the bid's initial commitment to one- has put the momentum achieved in 2017 in a coma that, as the years progress, becomes increasingly harder to reverse.

This paper is based on the official bid and evaluation documents, a series of interviews with key participants in the event (planners, administrators, artists, volunteers), media reports and extensive field studies during the bidding, planning, implementation and post-implementation stages. The study aims to contribute to the literature on the interaction between mega-events and small city urban development and provides insights for urban planners, city managers and public policy administrators on the significance of citizen participation in mega-event planning.

Participation as a key issue in mega-events

The idea that hosting a mega-event can activate urban regeneration through the spatial and temporal intensification of investment in infrastructure, innovative policies and global media exposure (Chalkley and Essex 1999; Hiller 2006; Gold and Gold 2008; Kassens-Noor 2012) is being steadily challenged. Organizing a mega-event is often burdened by budget and scheduling excesses, over-scaled and under-used structures, gentrification and social discrimination, environmental concerns, 'practices of exception' in urban governance and unconvincing legacy narratives (Agamben 2005; Baptista 2013; Müller 2015; Lauermann 2019). The scale of a mega-event, self-evidently large when it comes to participants, spatial infrastructure and financial recourses, seems to generate second thoughts for local governments who grow increasingly hesitant to undertake the political and economic risk of heavy investments with uncertain outcomes (Evans 2018; Zeghni and Fabry 2019) or are faced with urban opposition (Lauermann and Vogelpohl 2019). Furthermore, the links of mega-events with historic- and heritage-rich cities, already posing pressing practical issues, need further theoretical study (Jones & Ponzini 2018), in order to grasp the criticalities of spatializing a mega-event in places already laden with multiple meanings.

Citizen involvement from the earlier stages is suggested as one of the ways to mitigate the adverse effects of mega-events on cities and regions, without compromising their benefits, as they 'not only facilitate a better alignment of infrastructure with citizen needs but also build consensus and reduce potential opposition' (Müller 2015, 14). Leonardsen (2007) discusses the critical importance of stakeholder and public involvement in mega-event planning, while Kassens-Noor and Lauermann (2017) also make a case for the inclusion of a wide variety of stakeholders to prepare a realizable and effective plan, in terms of urban potential and legacies. The role of citizen participation in mega- or non-mega-events is extensively present in the literature. Regarding the contributions to the planning stage while focusing on the tourism aspect, Lamberti et al. (2011) analyze local involvement in the tourist policies of the 2010 Shanghai World Expo, and Pappas (2014) examines residents' views of the 2012 London Olympics in correlation to their participation in decision making on tourism development. Additionally, a study of Kaohsiung's residents' pre-event perceptions of the 2009 World Games in Taiwan (Ma et al. 2011), reveals the absence of community involvement from the planning process and the misalignment between local expectations and organizational aspirations. Public perception at the end of an event is showcased in a series of studies on the effects of a specific event on its audience (e.g. Ramchandani et al. 2015, on the impacts of sporting events on their audiences, or Lähdesmäki 2011, on the audiences' perception of the representations of identity aspects in ECoC projects), or the differences of an event's perception between engaged participants and more distanced audiences (Santos et al. 2017). Understandably, such studies have a very specific focus on the short- or long-term outcomes of particular participatory modes, or they zoom in on specific 'moments' within the expanded timeframe of an event. What appears to be missing is a wider-angle look at the various forms of participation and their evolution in the lifetime of an event, in order to better understand the intentions, successes and possible pitfalls of participatory processes. Involvement in collateral planning and consumption of an event as audience are the two ends in the wide range of participatory tropes regardless of event size. In cultural mega-events, participation can expand in other directions, the four main modes of which are as audience, as active co-creator of cultural content, as project or entire event designer, and as volunteer (Tommarchi, Hansen, and Bianchini 2018).

Participation in cultural activity legitimizes public spending, has a positive impact on social and individual problems (such as social isolation or low self-esteem), and appears to contribute to citizens' social and political engagement, which is particularly pertinent in a global climate of successive democratic crises and ambient distrust in politics. Therefore, although participation was not one of the initial objectives of the ECoC projects, it gradually emerged as one of the key criteria with which a city's ECoC application is assessed. Participation is deemed crucial both for the success and the legacy of the event (European Commission 2014, 20) and is consequently one of the main issues considered in the ex-post evaluation of an ECoC event (European Commission et al. 2018, 10–11). ECoC evaluations look at the audience numbers and their social characteristics, the volunteer programs and the community projects in order to gauge the degree of the event's outreach to the local and international community. Target participants involving schools, minorities or other less-advantaged groups are encouraged, even expected. Therefore, participation is not simply a matter of numbers, but an opportunity for power as it involves having a voice and a degree of influence in decision-making, be it politics and media (Carpentier 2015), urban planning (Arnstein 1969), or cultural production (Jancovich 2017). And, although many cities specifically target active citizen involvement in decision-making processes, this often trickles down to pursuing audience participation or may be regarded from a neo-liberal point of view as carrying the responsibilities of the state over to the citizens (Jancovich and Hansen 2018).

Participation has another side to it, forming the basis of bottom-up strategies, which can be particularly effective in the case of small cities. Such strategies can exploit a community's bonds, its comprehension of combined strengths and its understanding of collective needs and wishes. Although the spotlight mostly falls on the larger metropolises, small and medium-sized European cities and towns (SMSTs) are home to 38% of the European population (European Commission 2011, 2). Compared to great urban centers, SMSTs may generally exhibit economic and demographic stagnation, economic inefficiency and outmigration of smart, talented and knowledgeable people. They may also display inequities, in terms of social justice, economic opportunity and income, as well as overall lack of resources, environmental degradation, gentrification, neglect and encroachment of the cultural landscape. In exploring the main features of the three types of European SMSTs' socio-economic profiles, Hamdouch, Demaziere, and Banovac (2017) suggest that to shift developmental profiles from a residential economy towards competitiveness and innovation, towns could encourage the formation of networks and social capital in order to counter the drawing power of large cities and attract both residents and investment. In this direction, SMST-wide participation in a mega-event may spark the launching and generation of such networks. Knox and Mayer (2009) discuss the effectiveness of community-based approaches in enabling small cities to overcome their inherent disadvantages and achieve sustainability; in contrast to large metropolitan regions, small cities can utilize their potential of 'smallness' to employ creativity, heritage and culture as tools for urban regeneration, which can be implemented with greater attention to the community's needs, and spatially dispersed throughout the entire urban fabric and rural region. Moreover, community-based methodologies rely on the distinctiveness of smaller urban places -their history, heritage, traditional skills, past expertise and tightly woven community-, and build upon their collective creative expressions of culture and heritage to give meaning to their continued existence, preserve and further develop their identity and instigate vitality (Landry 2012). Focusing on the community's aspirations and needs, these bottom-up strategies employ art and culture to create vital opportunities for engagement among citizens, visitors, neighbors, friends and families. Simultaneously, they initiate public participation platforms for citizens to collaborate through varied and multifaceted leadership and establish social capacity practices that empower community members and build social capital (Cuesta, Gillespie, and Padraic 2005). Through art and culture and based on their community and heritage, residents and artists can visualize a shared future, upgrade their quality of life, re-establish a cultural identity, shift prevalent thinking, and consequently instigate social change, activate socio-economic revitalization and bring about civic renewal (Knox and Mayer 2009). It is with these thoughts in mind that we look at the participatory practices employed in the Pafos 2017 events, at how and to what extent Pafos exploited its community potential to bring about much-publicized change.

Pafos in a nutshell

In over 3,000 years, Pafos, on the west coast of Cyprus, has evolved from a small Mycenaean outpost to a celebrated Sanctuary of Aphrodite; from a Roman seaside capital to a provincial town on the hills; and from a fishing and agricultural past to a mass-tourism-dependent future. The city of Pafos, with a population of 32,892 in 2011 (Statistical Services 2019), is the administrative and financial center of the District of Pafos, which comprises an area of 1,393 km^2, with four Municipalities (Pafos, Geroskipou, Pegeia and Polis Chrysochous) and 98 village communities. Although it ran on behalf of the entire District, with a total population of 90,295 (in 2011), it is still one of the smallest cities to ever run –and eventually hold- the title of ECoC. Its diversity however is striking: almost 30,000 -about 1/3- of the district's residents are non-Cypriot: British, Russian, Greek, Romanian, Bulgarian, Scandinavian, who live and work in different parts of the district. Tourism is the engine that propels the region's economy: in 2010 Pafos offered about 29,000 beds (31% of the available beds in Cyprus), 40% of which in 4- and 5-star resorts (PCCI 2010, 18), and attracted almost 32% of visitor arrivals in the island. In 2013 the region of Pafos welcomed 930,886 visitors from all over the world, a number which rose to 1,365,500 in 2017 (Statistical Services 2020a) and brought €985 mil. in revenue (PCCI 2017, 17). Between 2010 and 2019, accommodation and food service activities provided 15%-20% of the jobs in Pafos, along with trade/services (17%-14%) (Statistical Services 2020b), while arts (bundled with entertainment) only a meager 2%. Construction is also booming, after a slump in 2013–15, offering 16% of the jobs in 2019, with real-estate sales rising; in 2017 Pafos and Limassol were the foreign buyers' favorite areas, together accounting for 70% of the total real-estate sales to foreigners (PCCI 2017, 22). These recent developments draw a much brighter picture than the one in 2011, when the District of Pafos had the highest ratio of unemployment in relation to its population in the Republic, at 6.9% (Statistical Services 2015).

Based on the above, Pafos's economy manifests most of the 'residential' economy characteristics (Hamdouch, Demaziere, and Banovac 2017), since it relies mostly on local activities that satisfy the needs of its residents and visitors, and its products are destined for local consumption. These numbers also clearly indicate two features of Pafos that distinguish it from other aspiring ECoCs: (i) the small size of its permanent population makes the existence of large-scale cultural infrastructure unsustainable, and (ii) as an already well-established tourist destination, Pafos was more interested in becoming a cultural destination, asking visitors, but mainly its residents, to re-engage with the city (Dova 2013).

When bidding for the 2017 European Capital of Culture in Cyprus, Pafos had very little to show on the cultural front. In 2010 Pafos had one theatre and one cinema, both inefficiently housed in adapted industrial spaces, and, despite a long-standing local interest in cultural activities, only a handful of galleries and artists' studios. Its two extensive archaeological areas, the urban ruins of Nea Pafos and the Tombs of the Kings necropolis, are the only breaks in the series of five-star resort hotels that line the Pafos seafront. Meanwhile, the haphazard expansion of the city after 1974[2], both along the coast and inland, had left behind the historic center of the city, its buildings and public spaces neglected and derelict. A distinctive characteristic of urban space in the Republic of Cyprus is the presence of Turkish-Cypriot (T/C) property, which, following the forced population relocation after 1974, is currently managed by a special branch of the Cypriot Ministry of Interior, the T/C Management Services (Legislation 139/91–68(I)/2012). This service controls the use of T/C property by displaced Greek-Cypriots under favorable conditions, in order to alleviate the problems caused by the 1974 events. The center of Pafos has more T/C properties than any other city in the Republic (Phedonos 2019). With many of them left empty or deteriorating, Pafos's urban core was both an obvious case of urban decay and a strong reminder of the political wounds of the island.

The 'Open Air Factory'; from build-up to closure

The 'Open Air Factory' in plan; joining forces on the basis of heritage

The Municipality of Pafos initiated the bidding process in Dec. 2010 with an open rally for citizens to support the candidacy of the city (Pafos Working Group 2012, 10). The effort was continued through a series of meetings, the Workshops of Ideas, in which 200 people from the locality were invited to participate (public servants, academics, cultural associations, artists and marketing experts). The Pafos 2017 Office for the ECoC candidacy was set up by the Pafos Municipalities' Committee in Feb. 2011, comprising municipal servants, an artistic director, management and communications experts, and invited academics from the local Neapolis University (ibid., pp. 36-38). This Working Group (2017 WG), responsible for preparing the first application, sketched out the idea of the Open Air Factory to counterbalance Pafos's lack of indoor cultural infrastructure and to create an accessible framework that could reach out to the local population (Grizzo et al. 2011). Even from the pre-selection stage, the Selection Panel noted for Pafos the necessity to integrate a clearer community engagement strategy (Selection Panel 2011, 11); in the second bidding phase, the 2017 WG began to regard the ECoC application as a long-awaited platform for dialogue on the future of the city.

The participatory nature of the second bidding phase has a prominent position in the application documents (Grizzo et al. 2011). After the 2017 WG extended an open invitation to all Pafians (through newspapers, social media, word-to-mouth), direct public involvement in the pre-ECoC events (cultural events, seminars, information meetings) reached 8,000 people, an impressive number for the region's population, and 800 people registered as participants in the enlarged Workshops of Ideas. The latter were distributed into seven fields: Artistic Creation, Urban Development and Infrastructure, Volunteerism, Youth and Culture, Social Inclusion, Entrepreneurship and Culture, Marketing and Communication, each led by members of the WG, who coordinated the process. Establishing its headquarters in the District Officer's Residence, a prominent Pafos landmark, the WG made its presence a public affair, and its location a strong urban reference point. Meetings were expected to be held twice a month, but the urgency of the call and the participants' enthusiasm pushed for once- or twice-a-week meetings. These Workshops produced both the basis and the specifics for all aspects of the finalized ECoC bid. Their members were a cross-section of the local society: independent artists and representatives of local cultural groups (e.g. House of Letters and Arts, the Ex-Artis cultural initiative), professionals (e.g. architects, journalists, educators, business owners, hotel managers), students and representatives from diverse local associations like the Association of Ladies of Pafos, the Federation of Parents of Pafos, or the British Ex-Pat Association. It must be noted, however, that most of the participants were individual citizens with an active interest in culture, and not members of structured groups or stakeholders' representatives. This, we believe, provided much of the strength of the Pafos bid. The popularity of this process initiated another volunteers' program, the Direct Communication Team, with 200 members, to support the pre-ECoC events and liaise with the locals and visitors in Pafos. This comprehensive effort led to the general embracing of the 2017 Open Air Factory strategy by the people of Pafos, in a very short time after the pre-selection stage. It established a space where everyone was encouraged to express ideas and participate in the decision-making process, forged a shared urban consciousness among people of diverse backgrounds, built the feeling of joint ownership for the project and, most importantly, shaped a collective vision for the future of the city, instigating hope. This significant change in the spirit of the town and the engagement of the general public was noticed by the Selection Panel, who appreciated 'the active and encouraging participation of citizens from the broader Pafos district in different activities' and 'the already well-established citizen ownership of the project' in their selection of Pafos as 2017 ECoC (Selection Panel 2012, 6).

Definitions and re-definitions of heritage became the ground upon which the diverse social groups of the city came together. The application, as prepared by the 2017 WG and the Workshops of Ideas, gives a prominent role to both tangible and intangible heritage, setting it at the basis for the program's three main production lines ('Myth and Religion', 'World Travelers', 'Stages of the Future') and cultural projects. Understood as a composite of all ethnicities that have constituted Cypriot identity over the centuries, heritage is proposed as a tool for inclusion and integration, as a way to introduce the various ethnic groups of the city to one another and connect them into a more cohesive social entity: ' ... to enrich Pafos's cultural heritage through modern cultural and artistic interventions and tie art and heritage in with an interactive and participatory approach for the citizens of all ages and ethnic groups' (Pafos Working Group 2012, 12). In this capacity, heritage is presented as the occasion for which the political and social tensions troubling the island could be alleviated: ' ... a reuniting of Greek Cypriots and Turkish-Cypriots promoted through The Open Khan multicultural center and The Big Mosaic processes of discovering a common heritage and traditions' (Pafos Working Group 2012, 7). The cultural projects included in the application were conceived on the basis of an all-inclusive approach through an expanded network of mini-labs composed of citizens, students and artists (Pafos Working Group 2012, 44–45). This process would be continued during the implementation of the event, in order to also support knowledge creation and networking among locals, and to further the integration of migrant ethnic and foreign resident communities. Participation on and through the creation of a shared heritage was therewith placed at the core of the application program.

The 'Open Air Factory' concept, inspired by the ancient tradition of culture in open-air public spaces, was subsequently developed as an idea that describes both the outdoor locations for most of the program and its public, inclusive nature as a site of cultural production that has room for everyone (Grizzo et al. 2011, 25–26). The Urban Development and Infrastructure Workshop focused on the regeneration of the city center in order to provide a shared civic space and a point of reference for the social groups of Pafos, and a counterpoint of urban character, culture and sociability to the banality of waterfront resorts. The particular character of the Pafos center and its Turkish-Cypriot properties, in terms of heritage, renders it as not simply a record of the city's urban development or a log of its architectural styles, but also as the bearer of a consonant symbiosis and a dissonant history, the reminder of both pleasant interaction and bitter conflict. Therefore, using this space for cultural purposes came to the forefront of discussions, also for its symbolic connotations. The team's idea was to use key buildings and public spaces in the city center and weave a network of indoor and outdoor spaces where cultural activities could take place (Dova, Sivitanidou, and Koutsolambros 2015). Hence, instead of spending large sums of money on infrastructure more fitting to a larger metropolitan region, many smaller venues could be remodeled, together with versatile open-air spaces, in order to hold both large and smaller audiences. The plan highlighted heritage sites, which until then had been underused or endangered, and integrated municipal projects in various stages of maturity, to strengthen this proposed urban fabric (Fig. 1).

The involvement of the community before, during, and after the ECoC events was to be assessed through the Pafos 2017 Barometer which was designed as the primary evaluation tool (Pafos Working Group 2012, 81–88) in order to aid the development of the

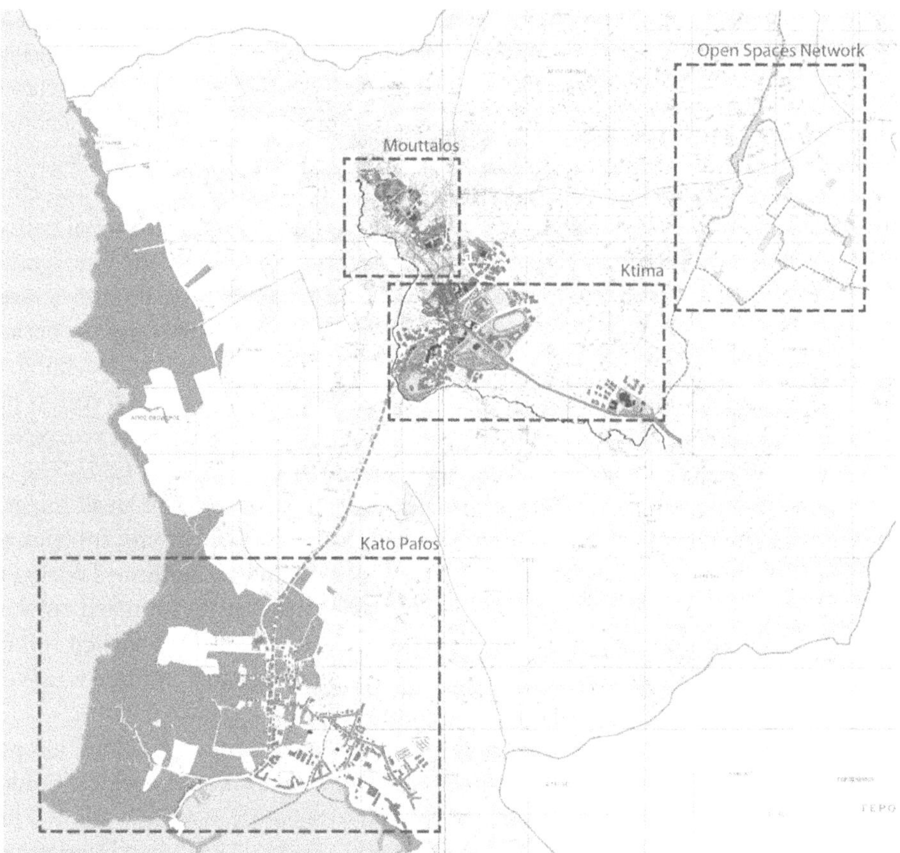

Figure 1. Map of projects proposed for the 2017 ECoC in the Pafos bid book (source: E. Dova, A. Sivitanidou, 2013).

district's cultural strategy and investment. The smallness of the district was expected to facilitate the research because results could be promptly collected and analyzed and immediately facilitate further changes in the process. The Barometer practicalities were extensively described in the bid: workshops for its development, diagnostic tools -from conventional questionnaires to digital games-, supervising and monitoring bodies, and synchronous communication of the outcomes. An ambitious plan, the Pafos 2017 Barometer was presented as the precursor of an island-wide cultural observatory system.

In correlating the fourfold of participation modes (Tommarchi, Hansen, and Bianchini 2018) with the bidding stage we can identify all four: the local community was directly and actively involved in the design of the entire project, in the creation of discrete cultural components, enthusiastic volunteers built the support of the wider community and convinced the selection experts, and audiences attended the preliminary events fleshing out the 'Open Air Factory' concept.

The 'Open Air Factory' at work; depending on the locals

The 2012–13 Cypriot financial crisis coincided with Pafos's selection as the 2017 ECoC, and it severely affected the city in general and the ECoC in particular, as the initial government funding, private sponsoring and municipal/community contribution were drastically reduced or difficult to secure: the ECoC budget was cut from €23 mil. to just €8.5 mil. (European Commission et al. 2018, 58). An additional major setback was the long delay in establishing the Pafos 2017 Organization due to political controversies (Mavros 2012; Nanos 2013), which pushed the timeframe of realization even further back. Although stated in the ECoC application, the Citizen Advisors body who would communicate with the Pafos 2017 Board of Directors on behalf of the citizens (Pafos Working Group 2012, 49) was never established, and there was no formal body to incorporate locals in the decision-making process. After the budget cuts, a very limited amount was allocated to community involvement and outreach activities (European Commission et al. 2018, 58). So, participation, instead of a process directly influencing the main program of events, became a separate program thread: the Community Involvement Program (CIP) that involved schools, minority groups and local artistic associations. As a detached part of the program, community involvement was restricted to discrete actions and no longer integrated into the overall design of the ECoC event. This separation is indicative of a notable departure from the application, one of many. Out of the 168 separate cultural projects realized in 2017, only 25 originated from the collectively designed bid (e.g. The Big Mosaic, Travelling Playground, Spatial Counterpoints, see: Doetzer 2019, 252), yet with a much-reduced scale and scope (due to budget and timeframe restrictions). An open call in 2014 for project proposals resulted in 30 proposals being realized until 2016, and 19 integrated in the 2017 ECoC, yet it is not clear how many of them were locally sourced. The reduced budget, however, limited the possibilities for many international projects, and working with local artists became necessary. The silver lining was that the event acquired a very strong Cypriot and Pafian flavor: local artists participated in 44% of the projects, while 60% of all projects were produced in Pafos (ibid., 256). The bid's strong emphasis on local heritage was evident in the final program of events: out of the 168 projects, 86 (55%) were immediately related to heritage[3], 138 (88%) were located in designated heritage sites, and 79 (50%) concerned heritage and took place in designated heritage areas. The majority was concentrated in the city of Pafos, but many of them were spread among the other towns and villages of the District (Fig. 2). As a Pafos 2017 Organization employee put it, 'notions of heritage were being completely intertwined in the spirit and fully embedded in the concept of the 'Open Factory', the design of its thematic categories and its implementation' (Dova et al. 2020, 278).

Audience participation was remarkable: the total number of audiences for the ECoC events in Pafos is 207,250 (European Commission et al. 2018, 81), which may not seem much in the large context of the ECoC program, but it is quite impressive for a city the size of Pafos, and its minimal experience in large-scale cultural events. It must also be noted that the actual number may be much higher, since this is an estimate based on the ticketed events, which constituted only 20% of the entire project (Doetzer 2019, 255). Faithful to the principles of the 'Open Air Factory', most of the projects (60-70%) were indeed delivered in open-air locations (European Commission et al. 2018, 62), spreading the ECoC activities in the city and the region, and reaching wider audiences; people often ran into events by chance, which inserted cultural activity into the everyday life of the locals. Mass participation activities were favored by the organizers, and, since many activities did not require tickets or pre-planning, they were more accessible and inclusive. The spread of the activities throughout the year -thanks to the Pafos weather- also made it easier for audiences to participate in many events rather than having to choose from a densely populated cultural calendar (ibid.). A number of

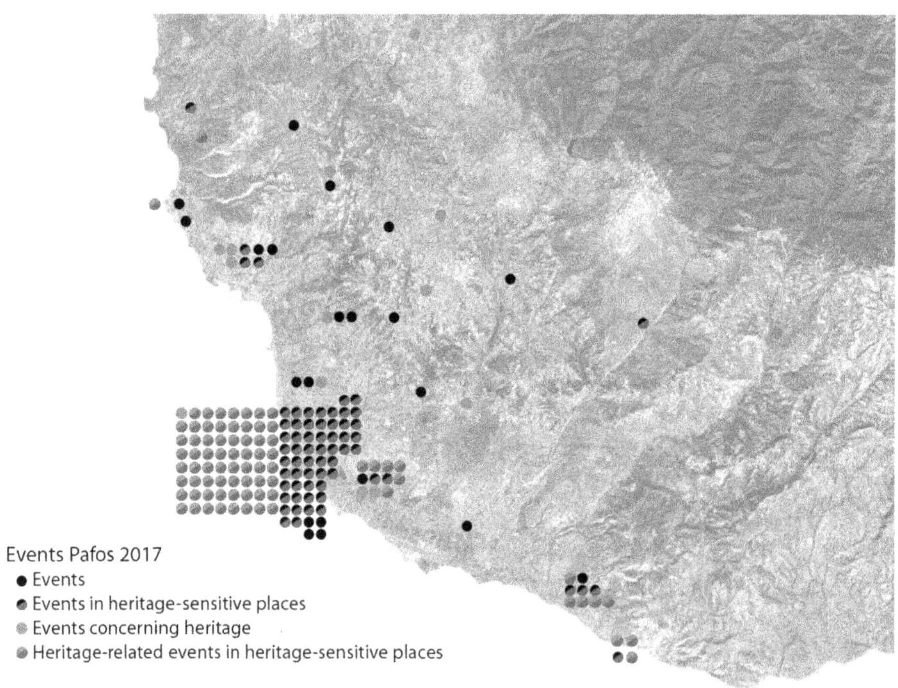

Figure 2. Map of the District of Pafos, showing the location of events and their relationship to heritage (source: A. Sivitanidou, 2018).

activities were located in lesser-known areas and villages of the Pafos district, giving locals the opportunity to discover locations they didn't know existed (e.g. 'Tours and Traces', visiting Androlykou, Panayia tou Sinti, etc.). The dearth of previous evidence makes it impossible to trace the changes in the relationship of locals to cultural events, but the public consultation conducted for the Pafos 2017 Barometer shows that about 80% of the respondents were encouraged to attend more cultural events, and a similar percentage (84%) would attend a wider range of events. 67% of the respondents also felt that the ECoC had introduced more people to culture, and almost 60% attended more cultural events as the year progressed. The small budget affected the impact of the projects on specific community groups, and it appears that some were left behind; the Ex-post evaluation of Pafos 2017 observes that no projects specifically targeted disadvantaged groups, and many tolerance-themed events were too small to address larger groups, focusing rather on individual perceptions (European Commission et al. 2018, 61, 82). The scope of the projects was too small to reach citizens residing in remote rural communities and outside the Pafos district, expats and Turkish-Cypriots, who did not have equal opportunities to participate due to communication difficulties, long distances and restricted resources.

Volunteerism was the area through which locals felt increasingly involved with the 2017 ECoC, and appears to be one of the most socially penetrating aspects of the Pafos ECoC. In 168 events, Pafos 2017 mobilized about 350 volunteers of different ages (14-77), professions and backgrounds (European Commission et al. 2018, 82), of which 83% Cypriot and 17% of other nationalities (Pafos Volunteer Office 2018, 3), with a total of 2612 presences. Again, one should look at the number in relation to the size of the city, but also in relation to the city's past volunteering efforts, which were limited among small circles (e.g. within Ex-pat associations, Ladies' Circles and Animal Protection groups, boy scouts, all too small to provide relative data). Apart from an obvious mode of citizen engagement with the event, volunteerism was also a means to overcome the budget limitations; the Pafos 2017 Organization being staffed with a personnel of 17 (when other ECoC secretariats are between 30-50), the presence of volunteers was not simply symbolic of the city's dedication, but also crucial to the operational capacity of the Organization. Volunteers received training and their responsibilities spread from technical and organizational support to the active design, governance, and even the material construction of certain projects (European Commission et al. 2018, 72). Even the Board of Directors, whose members were local professionals, entrepreneurs and politicians, provided their services voluntarily, as a further manifestation of the city's commitment to the ECoC project that, by 2017, had become a matter of local pride.

The remodeling of the city center greatly contributed to widespread feelings of city renewal (Fig. 3). Although the proposed Pafos 2017 infrastructure plan was not included in the ECoC budget, it provided the Municipality of Pafos with a 'master plan' for urban remodeling. After the city's designation as ECoC, the Municipality had both the motive and the obligation to realize it, along with a justification to apply political pressure. Therefore, it developed the Integrated Sustainable Urban Development Plan (ISUDP 2014-2020; Deloitte 2013) in order to claim assistance from the EU Cohesion Fund for Investment in Growth and Employment, targeting (i) the protection and promotion of cultural heritage, (ii) the enhancement of small and medium-sized enterprises'

Figure 3. The Pafos center with buildings and public spaces restored for Pafos 2017 (source: A. Sivitanidou, 2018).

competitiveness and (iii) the promotion of employment and the alleviation of social exclusion. Through the ISUDP, heritage preservation was directly linked with the urban regeneration program and indirectly related to quality of place and livability, job creation, financial viability and social cohesion (Dova, Sivitanidou, and Balasis 2019). By redefining the city center as a destination and a counterpoint to the waterfront, the ISUDP also aimed to diversify Pafos's tourist offerings through culture and heritage, generating the previously non-existent 'creative economies' (Georgi, Vissilia, and Koutsolambros 2014). Although little was finished at the beginning of 2017, as the year went on, the remodeling projects were being completed and used as ECoC venues (Fig. 4). Other historic buildings and archaeological sites of the city and the district (e.g. the Roman Odeon, the District officer's Residence, the Shelley Residence) hosted performances and exhibitions, for larger or smaller audiences, true to the concept of the Open Air Factory. Even less prominent buildings got their due: the *Peri Poleos* (About the City) project, curated by a local artist, gathered citizens' stories and presented the city's old patisseries, cinemas, legendary cafés and private houses to the public. *Spatial Counterpoints*, also a local production through an international cooperation, guided its audience in the remodeled center with a choreography inspired by the urban past. In these and many similar projects, intangible local heritage was manifested through the

Figure 4. The opening of the Producing Lines exhibition in the Attikon/ Othellos theater in Sept. 2017 (source: E. Dova, 2017).

tangible physicality of an urban space that was brought out of the shadows. It is not surprising, then, that the majority of interviewees (91.2%) believed that the ECoC would have long-term benefits to the city and citizens, namely urban revitalization, cultural development, and increase in tourism (Tsagkaridis 2018).

To sum up, it appears that although community participation in the overall design of the 2017 event was sidelined, the involvement of locals in content creation *and* consumption filled the gap, while the city's palpably growing transformation strengthened local understanding and acceptance of the Open Air Factory's objectives and impact.

The 'Open Air Factory' shut; its 'workers' left to their own devices

On December 14, 2017, a public meeting was initiated by the Pafos House of Letters and Arts, a local cultural organization, on the future of the ECoC project in Pafos. The year was closing but no succession plan had been announced, and concerned citizens were anxious to sustain the Pafos2017 legacy. Although the meeting had no clear result, the massive participation (local entrepreneurs and professionals, academics, central and local administration, artists, students), the statements of the local cultural organizations' representatives and the subsequent discussion made evident the community's acknowledgement of the ECoC's positive impact on the cultural capacity of the city, on its physical space, and also on people's civic pride, and its agony to sustain the 2017 legacy (Nomisma 2017). This meeting is also indicative of the local expectations for the future of Pafos after the festivities' end, and of the bottom-up spirit built within the ECoC effort. The Pafos 2017 Organization, however, was dismantled in early 2018, and despite the commitment in the bid and repeated press announcements by the Municipality, the Organization of

Cultural Development of Pafos (the body that would replace Pafos 2017) is still in the planning stage (Nanos 2019). The ten-fold increase in cultural events in Pafos between 2016 and 2017 (European Commission et al. 2018, 80) seemed to collapse in 2018, with only a few local groups (e.g. Ex-Artis, Kimonos Centre, Pafos Photographic Society, Folklore Group Ktima) managing to hold cultural events thanks to their gained expertise, newly-established networks and acquired confidence. Yet all these groups had to work on their own, without a coordinating mechanism to support planning, financing, communication, etc.[4] All in all, cultural activity in the city is being sustained mostly at an ad-hoc level through isolated instances and personal drives.

This section is painfully short, because the end of 2017 signaled an abrupt break in the city's cultural activity. Although the spatial projects are only indirectly associated with the Pafos 2017 Organization, the 2017 legacy appears to be literally located in the urban space of the city, whose remodeling is still in progress, as the Municipality has extended the scope of the intervention beyond the initial 2017 plans (e.g. the Laiki Geitonia -Traditional Neighborhood- project added to the pedestrian network, the restored Municipal Market building, the Post Office, the Pafos District Administration Offices and their immediate environs, completed in 2018-2019). For the small scale of the city, the wide network of spatial interventions has made a big difference in the look, feel and function of the central urban area. For the area's residents, the reclaimed public places have improved the quality of life in the city center, and the growth of existing businesses is observable (Dova, Sivitanidou, and Balasis 2019). It is, of course, hard to attribute an economic boost to the urban remodeling, since there are yet no official reports or indicators for Pafos regarding job creation, resident attraction, tourist arrivals or additional production after 2017.

On the contrary, without the 2017 crowds, the city center seems remarkably empty. Despite the improved cultural infrastructure, the absence of a cultural programming/management body, which was previously discussed in this article, is negatively affecting the city's attempts to develop the creative economies sector. As with almost all of the spatial interventions envisioned in the ECoC infrastructure plan and implemented through the ISUDP, the Ibrahim Khan is a good example of a project with a promising beginning but an unclear future. A flagship 2017 project, symbolic of Pafos's multi-cultural past, it was remodeled as a cultural venue, and hosted many ECoC projects, but its spaces were filled in 2018 by small artisanal and specialties shops. However, the Khan's full potential as a cultural venue has not been exploited, as the library and the artists' residency program proposed in the Pafos 2017 bid are not operating, while the gallery and the open-air theater are underused. In the absence of a longer-term managing scheme, it is the shop owners who have been active in reanimating the Khan and are hosting mini-events (open houses, concerts, festivals, Christmas markets, etc.) to invite the public. In this and similar cases, the small scale of the city has an adverse effect, as it is difficult to sustain cultural and creative production without the constant input and coordination provided by a dedicated agency.

Conclusions: A mixed outcome

For the small city of Pafos, community participation became a feasible strategy and played an integral role in the application for the 2017 ECoC and its implementation.

Pafos' scale and under-developed cultural sector facilitated -in retrospect, it required- the active involvement of a large number of volunteers in all projects, and it eased increased levels of audience participation among locals. When the 2017 bid provided a structured forum for the development of ideas, the local community was eager to participate, and the perceived potential for realization through the ECoC nourished the effort. The extended family- and close personal relationships prevalent in the Pafos society may create close-knit and distinct social groups, but in the case of Pafos 2017 they managed to pull the community together and spread its ideas widely, through the volunteer groups, without requiring specialized resources like community meetings through representatives, or extensive and time-consuming surveys. Also, the area's limited artistic expertise and cultural infrastructure left room for the innovative exploitation of the area's abundant resources in nature, history, tangible and intangible heritage, as exemplified in the 'Open Air Factory' concept. The lower complexity of governance in a small-scale city like Pafos allowed for more immediate expression of ideas and concerns to the Municipal administration, but on the flip side, it seems to create insurmountable difficulties in continuing the Pafos 2017 legacy. The proposal for a new cultural carrier, suspected as a political maneuver, failed to inspire the general public, artistic organizations and individual artists in contributing or participating. Additionally, while the Pafos 2017 effort was characterized by adhocracy and transparency, these new efforts appear as rather esoteric and riddled with bureaucracy.

The end of 2017 in Pafos was met with ambivalence. On the one hand, Pafos had indeed managed to 'put together a suitable program to be a European Capital of Culture', and the entire city had been actively involved in the festivities, proving that 'an ECoC can be small but still successful and an ECoC is not always about putting on the 'biggest' and most expensive show possible' (European Commission et al. 2018, 88). It was due to its openness and inclusivity that the ECoC event was whole-heartily embraced and utilized by locals, leading to a strong willingness and new-found confidence to participate as active citizens and volunteers, artists and creators, working on cultural projects either individually or in groups. As such, the objective of building the cultural capacity of the area, i.e. providing venues, improving the skills of the cultural sector and contributing to cultural development, adding to the socio-cultural framework of the city, raising awareness, increasing participation and establishing an audience, was achieved. The urban center was also drastically transformed, as a result of a collective plan which aimed to highlight and reuse shared urban heritage. In these terms, the 2017 ECoC in Pafos could be considered a success.

On the other hand, little has been done in the years following 2017 to take advantage of this potential. Despite its mention in the bid, no succession plan was prepared, and opinions about the management of cultural venues and activities are still diverging. As the ECoC evaluation notes, the problem of the ECoC legacy is 'exacerbated in Pafos because the cultural sector does not have enough scale to automatically 'take on' new and additional cultural projects without the injection of funding and interest an ECoC year brings … and extra effort is required to ensure legacy is given its fullest chance possible to succeed' (European Commission et al. 2018, 88). Additionally, the urban remodeling has not yet contributed to increased cultural or financial outcomes; the lack of a legacy plan, combined with the small size of Pafos that cannot sustain cultural activity on 'autopilot', have left these reclaimed spaces largely empty of activity (Fig. 5). A lesser-noted impact is on tourism, which, without the supportive framework of 2017, has relapsed back to the conventional sun-and-sea model, despite its short-lived shift towards a more sustainable model of culture-sensitive tourism during the ECoC year.

The extent to which the citizens' involvement in the Pafos 2017 ECoC has affected their own perceptions of the city they live in, both immediately after 2017 and in the following years -an explicit goal of the ECoC bid-, still needs to be studied, as well as the degree of involvement versus mere attendance, residents' points of view and perceived attributes, and also how all these relate to the enhancement of the city's image (e.g. Santos et al.). Similar to investigations on volunteering in sport mega-events (e.g. MacLean and Hamm 2007; Holmes, Nichols, and Ralston 2018; Kim et al. 2019), future research should also flesh out the volunteers' views on their involvement in co-creating, producing and managing cultural events. Furthermore, and taking into account the significance of tourism for Pafos and Cyprus in general, the willingness of citizens to participate in future mega-events, and its implications for tourism planning is another direction to pursue.

Urban scale, in the case of Pafos, has been a mixed blessing. It facilitated the involvement of the local community in a variety of roles and made the success of the event a matter of collective responsibility, joined ownership and civic pride. Bottom-up participation and capacity-building processes supplied knowledge and experience, and set heritage –the heritage of the many and the heritages of the few- at the base of both the cultural and the spatial program of the 2017 ECoC. Though the smallness of the city served as a collective fabric for social relationships in 2017, once the driving force of

a b c

Figure 5. Left and center: The Ibrahim Khan after its remodeling in 2017. Right: The Attikon/ Othellos Theater during works in 2017 (source: E. Dova, 2018).

the ECoC was removed, the cohesion of the various socio-cultural strata was also lost. What Pafos 2017 was extremely successful in was raising expectations about what a small city in a time of crisis could accomplish through culture and heritage: a feat that turned to disappointment by the abrupt stop of cultural activity in 2018. A significant lesson learned from the Pafos case and its inability to capitalize on the community-driven success of the ECoC involves the commitment that should be expected, especially from small cities, in setting forth mechanisms and clauses for a legacy plan. The Pafos example makes the case for clear legacy directives and specific action plans to be required as early as the bidding process and carry substantial weight on the host selection. Lastly, the inclusion of community involvement practices and bottom-up strategies not only whilst planning, organizing and implementing cultural mega-events, but also when setting the legacy plan, could strengthen the longevity of the mega-event's positive effects and help ameliorate future tensions in the strenuous relationship between heritage and development.

Notes

1. With an urban population of almost 33,000 (in the most recent census of 2011), in the scale of the European context Pafos falls within the small-medium sized town (SMST) category, with populations of between 5,000 and 100,000 inhabitants (Servillo, Atkinson, and Hamdouch 2017). Given, however, that the entire population of the Republic of Cyprus does not exceed 900,000, and that Pafos is the country's fourth largest urban settlement, it is considered a city for the scale of Cyprus.
2. After the violent events of 1974, thousands of Greek-Cypriots were forcibly relocated from their homes in the north to the south, whereas Turkish-Cypriots had to take the reverse route. Cyprus has since been divided into the Republic of Cyprus (which controls the southern two-thirds of the island) and the self-declared Turkish Republic of Northern Cyprus (which controls the northern one-third). This displacement has emptied out large urban areas and entire villages all over Cyprus, while new neighborhoods and ad-hoc settlements sprang up around existing communities.
3. Authors' estimate based on the description of projects (Grizzo et al. 2011).
4. Pafos 2017 set aside approximately €500,000 for cultural activities after 2017 and the Ministry for Education and Culture offered an additional €270,000 to continue the Pafos 2017 legacy (European Commission et al. 2018, 86). However, without an established administrative body, these funds are inaccessible.

Acknowledgements

Sections of this paper are based on and supported by the research conducted within the JPI Cultural Heritage research project 'HOMEE – Heritage Opportunities/ Threats within Mega-Events in Europe', with a grant issued by the Cyprus Research Promotion Foundation (RESTART 2016-2020, Project: P2P/JPICH_HCE/0917/008). The authors would also like to thank Georgia Doetzer, Chief Executive at Rialto Theatre, Limassol, and Artistic Program Director of the Pafos 2017 Organization, for her input on the Pafos 2017 volunteers. The valuable comments of the editor, Davide Ponzini, and the anonymous reviewers, on an earlier draft of this paper, are greatly appreciated. Any errors or oversights are the authors' responsibility.

Disclosure statement

No potential conflict of interest was reported by the author(s).

References

Agamben, G. 2005. *State of Exception (K. Attell, Trans.)*. Chicago: University of Chicago Press.

Arnstein, S. 1969. “A Ladder of Citizen Participation.” *Journal of the American Institute of Planners* 35 (4): 216–224. doi.10.1080/01944366908977225.

Baptista, I. 2013. “Practices of Exception in Urban Governance: Reconfiguring Power Inside the State.” *Urban Studies* 50 (1): 39–54. doi.10.1177/0042098012453858.

Carpentier, N. 2015. “Differentiation Between Access, Interaction and Participation.” *Conjunctions: Transdisciplinary Journal of Cultural Participation* 2 (2): 9–28.

Chalkley, B., and S. Essex. 1999. “Urban Development Through Hosting International Events: A History of the Olympic Games.” *Planning Perspectives* 14 (4): 369–394. doi.10.1080/026654399364184.

Cuesta, C., D. Gillespie, and L. Padraic. 2005. *Bright stars: Charting the impact of the arts in rural Minnesota*. Accessed January 21 2020. https://www.mcknight.org/system/asset/document/76/pdf-6-mb.pdf

Deloitte. 2013. *Outline of Strategic Goals and Actions Relating to the Integrated Sustainable Urban Development (ISUDP) of the Municipality of Pafos in the Framework of the Planning Period 2014-2020*. Pafos: Pafos Municipality.

Doetzer, G. 2019. “Culture at the Core of the City’s Development. Pafos, European Capital of Culture 2017.” In *Transforming Cities Paradigms and Potentials of Urban Development Within the ‘European Capital of Culture*, edited by W. Schneider, and K. Jacobsen, 249–258. Hildesheim: Georg Olms Verlag.

Dova, E. 2013. “Organizing Public Space as a Framework for re-Thinking the City: Pafos’s bid for the 2017 European Capital of Culture.” In *Proceedings of the International Conference on “Changing Cities” (CD-ROM)*, edited by A. Gospodini, 2521–2530 Thessaloniki: Grafima Publications.

Dova, E., A. Sivitanidou, and E. Balasis. 2019. “Grafting Culture Onto Urban Space: The Exploration of Community-Based Practices Towards Sustainability in the Case of Pafos, Cyprus.” In *Tangible – Intangible Heritage(S) – Design, Social and Cultural Critiques on the Past, Present and the Future, Vol. 1*, edited by M. A. Segantini, 290–301. London: AMPS Research Organization.

Dova, E., A. Sivitanidou, J. Georgi, N. Tzortzi, and N. Anastasi. 2020. “Pafos 2017 European Capital of Culture. Motivating new Planning Practices and Rethinking Heritage.” In *Mega-events and Heritage: the Experience of Five European Cities*, edited by D. Ponzini, F. Bianchini, J. Georgi, N. Tzortzi, and J. Sanetra-Szeliga, 256–315. Krakow: International Cultural Centre.

Dova, E., A. Sivitanidou, and Y. Koutsolambros. 2015. “Experiments in Place-Making: Re-Thinking Pafos for 2017.” In *International Conference BioCultural 2015: Sustainability in Architectural Design Heritage Proceedings, 11-12 December, 2015, Vol. 1*, edited by M. Philokyprou, A. Michael, and A. Savvides, 184–193. Nicosia: Biocultural Research Programme, University of Cyprus.

European Commission. 2011. *Cities of Tomorrow: Challenges, Visions, Ways Forward*. Luxembourg: Publications Office of the European Union.

European Commission. 2014. *European Capitals of Culture 2020-2033 Guide for cities preparing to bid*. European Commission Culture. Accessed October 20 2019. http://ec.europa.eu/programmes/creative-europe/actions/documents/ecoc-candidates-guide_en.pdf

European Commission, Ecorys, & Centre for Strategy and Evaluation Services. 2018. *Ex-post Evaluation of the 2017 European Capitals of Culture*. European Commission, Directorate-General for Education, Youth, Sport and Culture. Accessed October 20 2019. https://ec.europa.eu/programmes/creative-europe/sites/creative-europe/files/ecoc_2017_ex_post_evaluation_final_report.pdf

Evans, G. 2018, November 18. Olympic Games: Why cities are snubbing the ‘greatest show on Earth’. *BBC News*. Accessed October 20 2019. https://www.bbc.com/news/world-46236682

Georgi, J., A. Vissilia, and I. Koutsolambros. 2014. "Pafos Cultural Capital of Europe 2017: Landscape Interventions in the Urban Fabric as a Tourist Development Tool." In *IFLA 2014, 51st World Congress of the International Federation of Landscape Architects/Thinking and Action: Earth, Home, Landscape of Places, Buenos Aires, Argentina, 5-7 June 2014, IFLA 51st World Congress Proceedings*, edited by International Federation of Landscape Architects, 425–431. Buenos Aires: International Federation of Landscape Architects.

Gold, J., and M. Gold. 2008. "Olympic Cities: Regeneration, City Rebranding and Changing Urban Agendas." *Geography Compass* 2 (1): 300–318.

Grizzo, N., T. G. Filippidis, K. Christoforou, and A. Iacovides, eds. 2011. *Pafos 2017 Open Air Factory*. Pafos: Municipality of Pafos.

Hamdouch, A., C. Demaziere, and K. Banovac. 2017. "The Socio-Economic Profiles of Small and Medium-Sized Towns: Insights from European Case Studies: Profiles of Small and Medium-Sized Towns." *Tijdschrift Voor Economische En Sociale Geografie* 108 (4): 456–471. doi.10.1111/tesg.12254.

Hiller, H. 2006. "Post-event Outcomes and the Post-Modern Turn: The Olympics and Urban Transformations." *European Sport Management Quarterly* 6 (4): 317–332. doi.10.1080/16184740601154458.

Holmes, K., G. Nichols, and R. Ralston. 2018. "It's a Once-in-a-Lifetime Experience and Opportunity- Deal with it! Volunteer Perceptions of the Management of the Volunteer Experience at the London 2012 Olympic Games." *Event Management* 22 (3): 389–403. doi.10.3727/152599518X15252895715050.

Jancovich, L. 2017. "The Participation Myth." *International Journal of Cultural Policy* 23 (1): 107–121. doi.10.1080/10286632.2015.1027698.

Jancovich, L., and L. Hansen. 2018. "Rethinking Participation in the Aarhus as European Capital of Culture 2017 Project." *Cultural Trends* 27 (3): 173–186.

Jones, Z. M., and D. Ponzini. 2018. "Mega-events and the Preservation of Urban Heritage: Literature Gaps, Potential Overlaps, and a Call for Further Research." *Journal of Planning Literature* 33 (4): 433–450. doi.10.1177/0885412218779603.

Kassens-Noor, E. 2012. *Planning Olympic Legacies: Transport Dreams and Urban Realities*. London: Routledge.

Kassens-Noor, E., and J. Lauermann. 2017. "How to Bid Better for the Olympics: A Participatory Mega-Event Planning Strategy for Local Legacies." *Journal of the American Planning Association* 83 (4): 335–345. doi.10.1080/01944363.2017.1361857.

Kim, H., Y. Choe, D. Kim, and J. Kim. 2019. "For Sustainable Benefits and Legacies of Mega-Events: A Case Study of the 2018 PyeongChang Winter Olympics from the Perspective of the Volunteer Co-Creators." *Sustainability* 11 (9): 2473. doi.10.3390/su11092473.

Knox, P., and H. Mayer. 2009. *Small Town Sustainability. Economic, Social, and Environmental Innovation*. Basel: Birkhäuser Verlag AG.

Lamberti, L., G. Noci, J. Guo, and S. Zhu. 2011. "Mega-events as Drivers of Community Participation in Developing Countries: The Case of Shanghai World Expo." *Tourism Management* 32 (6): 1474–1483. doi.10.1016/j.tourman.2010.12.008.

Landry, C. 2012. *The Creative City: A Toolkit for Urban Innovators*. London: Earthscan.

Lauermann, J. 2019. "The Urban Politics of Mega-Events." *Environment and Society* 10 (1): 48–62. doi.10.3167/ares.2019.100104.

Lauermann, J., and A. Vogelpohl. 2019. "Fast Activism: Resisting Mobile Policies." *Antipode* 51 (4): 1231–1250. doi.10.1111/anti.12538.

Lähdesmäki, T. 2011. "Celebration of Locality, Regionality, National Culture or Europeanness? Notions on Area-Based Identities in Pécs2010." In *Culture in/and Crisis*, edited by L. I. Komlósi, and W. Coudenys, 122–137. Antwerp: University Network of the European Capitals of Culture.

Leonardsen, D. 2007. "Planning of Mega Events: Experiences and Lessons." *Planning Theory & Practice* 8 (1): 11–30. doi.10.1080/14649350601158105.

Ma, S.-C., D. Egan, I. Rotherham, and S.-M. Ma. 2011. "A Framework for Monitoring During the Planning Stage for a Sports Mega-Event." *Journal of Sustainable Tourism* 19 (1): 79–96. doi.10.1080/09669582.2010.502576.

MacLean, J., and S. Hamm. 2007. "Motivation, Commitment, and Intentions of Volunteers at a Large Canadian Sporting Event." *Leisure/Loisir* 31 (2): 523–556. doi.10.1080/14927713.2007.9651394.

Mavros, P. 2012, November 29. [The Germans return]. *InBusiness*. Accessed October 20 2019. https://www.sigmalive.com/inbusiness/opinions/external/18887/oi-germanoi-ksanarxontai

Müller, M. 2015. "The Mega-Event Syndrome: Why So Much Goes Wrong in Mega-Event Planning and What to Do About It." *Journal of the American Planning Association* 81 (1): 6–17. doi.10.1080/01944363.2015.1038292.

Nanos, K. 2013, January. [The dispute over Paphos 2017 is intensifying with interceptions, threats and presidential elections]. *Politis [Online]*. Accessed October 20 2019. http://parathyro.politis.com.cy/2013/01/2017-5/

Nanos, K. 2019, October 18. [An Independent Organization of Cultural Development in Pafos is Being Founded]. *Politis [Online]*. Accessed October 20 2019. https://politis.com.cy/politis-news/kypros/idryetai-anexartitos-foreas-politistikis-anaptyxis-stin-pafo/

Nomisma. 2017, December 15. [Discussion at the Pafos House for Letters and Arts after Pafos 2017]. *Nomisma.Com.Cy*. Accessed October 20 2019. https://nomisma.com.cy/index.php?id=29890

Pafos 2017 Volunteer Office. 2018. *[Final Report for the Volunteer Program]*. Unpublished manuscript.

Pafos 2017 Working Group. 2012. *Pafos 2017 Questions & Answers*. Pafos: Municipality of Pafos.

Pappas, N. 2014. "Hosting Mega Events: Londoners' Support of the 2012 Olympics." *Journal of Hospitality and Tourism Management* 21: 10–17. doi.10.1016/j.jhtm.2014.02.001.

PCCI. 2010. *[Annual Report 2010]*. Pafos Chamber of Commerce and Industry. Accessed October 20 2019. https://pcci.org.cy/wp-content/uploads/2015/02/EBE_Etisia_Ek8esi_2010.pdf

PCCI. 2017. *[Annual Report 2017]*. Pafos Chamber of Commerce and Industry. Accessed October 20 2019. https://pcci.org.cy/wp-content/uploads/2018/10/ETISIA-EKTHESI-FOR-WEB.pdf

Phedonos, P. 2019, September 4. [Phedonos: The Management of T/C properties is still suffering]. *Sigmalive.Com*. Accessed October 20 2019. https://www.sigmalive.com/news/opinions_sigmalive/562757/faidonos-i-diaxeirisi-ton-tk-periousion-eksakolouthei-na-pasxei

Ramchandani, G., L. Davies, R. Coleman, S. Shibli, and J. Bingham. 2015. "Limited or Lasting Legacy? The Effect of non-Mega Sport Event Attendance on Participation." *European Sport Management Quarterly* 15 (1): 93–110. doi.10.1080/16184742.2014.996583.

Santos, J., L. Vareiro, P. Remoaldo, and J. Cadima Ribeiro. 2017. "Cultural Mega-Events and the Enhancement of a City's Image: Differences Between Engaged Participants and Attendees." *Journal of Policy Research in Tourism, Leisure and Events* 9 (2): 129–151. doi.10.1080/19407963.2016.1157598.

Selection Panel. 2011. *Designation of the European Capital of Culture 2017. Pre-Selection report.* Accessed October 20 2019. https://ec.europa.eu/programmes/creative-europe/sites/creative-europe/files/files/ecoc-2017-preselection-report-cyprus_en.pdf

Selection Panel. 2012. *Selection of the European Capital of Culture for 2017 in Cyprus. Final Selection report.* Accessed October 20 2019. https://ec.europa.eu/programmes/creative-europe/sites/creative-europe/files/files/ecoc-2017-panel-report-cyprus_en.pdf

Servillo, L., R. Atkinson, and A. Hamdouch. 2017. "Small and Medium-Sized Towns in Europe: Conceptual, Methodological and Policy Issues." *Tijdschrift Voor Economische En Sociale Geografie* 108 (4): 365–379. doi.10.1111/tesg.12252.

Statistical Services. 2015. *[Population Census 2011. V. I: General Population Characteristics, Immigration and Labour Force]* (No. 21). Republic of Cyprus, Ministry of Finance. Accessed October 20 2019. https://www.mof.gov.cy/mof/cystat/statistics.nsf/index_en/index_en?OpenDocument

Statistical Services. 2019. *Cyprus in Figures. General Statistics* (No. 24; Series II). Nicosia: Republic of Cyprus, Press and Information Office for the Statistical Service.

Statistical Services. 2020a. *[Tourism Statistics, 1999-2018]*. Republic of Cyprus, Ministry of Finance. Accessed October 20 2019. https://www.mof.gov.cy/mof/cystat/statistics.nsf/index_en/index_en?OpenDocument

Statistical Services. 2020b. *Labor Force Survey. 1999*-2019. Republic of Cyprus, Ministry of Finance. Accessed October 20 2019. https://www.mof.gov.cy/mof/cystat/statistics.nsf/index_en/index_en?OpenDocument

Tommarchi, E., L. Hansen, and F. Bianchini. 2018. "Problematising the Question of Participation in Capitals of Culture." *Participations* 15 (2): 154–169.

Tsagkaridis, K. 2018. *Anafora Apotelesmaton tis Dietous Ereunas tou Politistikou Barometrou (2016-2017) gia tis Epiptoseis tis Politistikis Proteuousas stin Pafo: Epidraseis stin Politistiki, Koinoniki kai Oikonomiki Anaptyksi [Report of the Findings of the two-Year Research of the Cultural Barometer (2016-2017) Regarding the Effects of the ECoC on Pafos' Cultural and Socioeconomic Development]*. Pafos: Pafos 2017 Organization - Neapolis University Pafos. Accessed October 20 2019. https://hephaestus.nup.ac.cy/handle/11728/10849

Zeghni, S., and N. Fabry. 2019. "Why do Cities Withdraw from Hosting the Olympic Games?" In *Hosting the Olympic Games. Uncertainty, Debates and Controversy*, edited by P.-O. Schut, and M. Delaplace, 9–23. London: Routledge.

A heritage-inspired cultural mega-event in a stigmatized city: Hull UK City of Culture 2017

Enrico Tommarchi and Franco Bianchini

ABSTRACT
While the impacts of cultural mega-events on a city's projected image have been extensively explored, less attention has been devoted to their role in relation to territorial stigmatization. These events have, in some cases, engaged with issues of stigmatization, as happened in Glasgow (European City of Culture 1990) and Derry-Londonderry (UK City of Culture 2013). This paper explores how built heritage is mobilized through a cultural mega-event in relation to stigma, through a case study of the city of Hull, in Northern England. Despite being among the country's main ports until the 1930s, Hull suffered from steady structural socio-economic decline, in particular due to the demise of its fishing industry from the 1970s onwards. Hull UK City of Culture 2017 was understood as an opportunity to counter negative external perceptions. Heritage was mobilized to tell the story of the city, engage residents and visitors in cultural activities and boost civic pride. Events in 2017 and beyond promoted heritage learning in innovative ways, also through the spectacularisation of heritage spaces. However, the official evaluation of Hull 2017 shows how positive immediate results were later scaled down, suggesting that these events are not sufficient on their own to counter stigmatization.

Introduction

Mega-events, in particular those that do not necessarily require huge amounts of resources (Di Vita and Wilson 2020), are attractive to policy makers as they may offer national or international visibility, support urban regeneration, attract additional funding and inward investment. These events have been understood as a means to reprofile cities and as a '"quick-fix" solution' to problems of external image (Quinn 2005, 932). In this context, some cultural mega-events (Jones 2020) such as the European Capital of Culture were in part characterized by attempts to counteract negative external perceptions and preconceptions about the host city. For example, the European City of Culture 1990 challenged the established image of Glasgow as a city affected by deindustrialization and decline (García 2005), while the UK City of Culture 2013 in Derry-Londonderry explored post-conflict reconciliation between the Protestant and Catholic communities (Boland, Murtagh, and Shirlow 2019). Some of these schemes tend to mobilize local history and heritage as assets to relaunch – often declining – cities as cultural tourism destinations and/or to diversify local economies. However, overall little attention has been devoted by researchers to the relationships between cultural mega-events and the stigmatization of places.

This article makes use of the concept of territorial stigmatization – i.e. the stigmatization of a place and of those who inhabit it – to explore how heritage (in particular built heritage) was mobilized within the Hull UK City of Culture 2017 (hereinafter Hull 2017) programme in relation to negative external perceptions and prejudices about Hull as a declining port city. This research is part of the European research project Heritage Opportunities/threats within Mega-Events in Europe (HOMEE).

In the following sections, a brief account of the concept of territorial stigmatization is provided and is discussed in relation to policy makers' perceptions of the role of cultural mega-events as opportunities to counteract negative external perceptions of a city. The case of Hull 2017 is presented and commented upon with a focus on the role played by heritage as a means to respond to the long-standing stigmatization of Hull as a deindustrializing port city. Finally, the legacy of the UK City of Culture in Hull and the use of its heritage-inspired approach to culture and regeneration beyond 2017 are discussed in the context of the challenges that the city is facing at the beginning of the 2020s.

Stigmatized places and cultural mega-events

The concept of territorial stigmatization was developed in particular by Wacquant (2007; 2008), drawing on the work of Goffman on 'social stigma' (1963) – who had not considered place as a factor fuelling stigmatization – and on Bourdieu's theory of 'symbolic power' (1991). Wacquant explored territorial stigmatization in the context of urban marginality and ethnic and social segregation in housing estates and ghettoes in France and the US, blemished as dangerous and infamous places because of poverty, crime, alleged immorality and their ethnic and social composition.

Territorial stigma is closely related to 'the stain of poverty, subaltern ethnicity [...], degraded housing, imputed immorality, and street crime' (Wacquant, Slater, and Pereira 2014, 1273). However, preconceptions about a place may also be related to other factors, such as the dominant social status of residents of wealthy neighbourhoods (Walmsley 1988), or the association of leisure districts with stimulus or of tranquil suburban estates with ennui (Ley 1983). Territorial stigmatization is therefore connected to the dominant narrative of a place, as in the case of the conceptualization of Las Vegas as the 'Sin City' (Nedélec 2017), and can be based on – not necessarily prominent – aspects of its history and heritage.

Dean and Hastings (2000) highlight how poor reputation is the outcome of long-standing beliefs about a neighbourhood and those who live in it, and that this reputation may become a stigma. This makes residents of stigmatized places often feel angry, hurt and upset. With regard to low-income housing estates, Wacquant (2008, 173) observes that 'the social taint of living' in a place 'that has become closely associated with poverty [...] affects all realms of existence'. The association of residents with a stigmatized place negatively affects their job opportunities, social relationships and access to public services (Dean and Hastings 2000; Wacquant 2008; Slater and Anderson 2012).

Residents often feel that they are subjected to unfair treatment by the media (Dean and Hastings 2000), which in many cases provide partial and biased accounts of their neighbourhood or city.

Stigmatization and urban regeneration

Territorial stigmatization has been connected with urban regeneration, gentrification and displacement. In their analysis of St. Paul's, in Bristol, Slater and Anderson (2012) show how renewal schemes for the area levered on its stigma to pursue gentrification. Paton, Mooney, and McKee (2012) suggest that urban renewal schemes and state-led gentrification may contribute to further pathologising areas that are perceived as 'problem places'. Kallin and Slater (2014) explore how the active blemishing of working class neighbourhoods may provide the rationale for state-driven extensive demolition and urban regeneration schemes tailored to attract wealthier households. Kirkness and Tijé-Dra (2017) suggest that policy makers may try to profit from stigmatization by exploiting the 'extraterritoriality' of these places or may take impulsive actions to attempt to eradicate the stigma. Speculative practices by real estate developers and financial actors may contribute to reinforce stigmatization, which can lead to demolition of affordable housing, displacement and renewal schemes targeted at wealthier city residents.

Event-led urban regeneration has been used in similar ways. While the impact of mega-events on the internal and external images of host cities has been documented (e.g. Richards and Wilson 2004), other studies have critically assessed mega-events' links with urban regeneration (Eisinger 2000; Mooney 2004). Paton, Mooney, and McKee (2012), for example, analyse the 2014 Commonwealth Games in Glasgow as a means to redefine working class people's lives in some of the city's deprived areas. Glasgow's East End was actively stigmatized in policy and media discourses to support the event as a means to pursue gentrification and capital accumulation (Paton 2018).

Heritage and urban regeneration

In this context, heritage plays a key role in transforming stigmatized places (e.g. De Cesari and Dimova 2019). However, as observed by Hewison (1987), heritage is mobilized predominantly in relation to nostalgia. This is in part due to the fact that nostalgia is in many cases an essential ingredient of tourism (Chhabra, Healy, and Sills 2003) which many urban regeneration schemes rely on. As a consequence, heritage narratives often display 'selectively embellished' accounts (Bélanger 2000, 386–387), which can fuel feelings of expropriation among the local community. Aspects of local heritage that are contested or problematic – i.e. 'dissonant' heritage (Tunbridge and Ashworth 1996) – are therefore largely neglected. The spectacularisation of urban space, heritage and culture is understood as a means to encourage cultural consumption (Zukin 1995), tourism and investment and circulate positive media narratives (Ponzini et al. 2019).

Hull UK City of Culture 2017

The case of Hull UK City of Culture 2017 was analysed as part of the HOMEE research project, focusing on the relationships between mega-events and built heritage. Initial desk research was followed by the analysis of relevant policy documents, including the event's bid (Hull UK City of Culture 2017, 2013) and final programme, the 2013 City Plan (Hull City Council n.d.), the Public Realm Improvement Strategy (Hull City Council 2014) and the city's Cultural Strategy 2016–2026 (Hull City Council 2016). The study also made use of the event's official evaluation reports (Culture, Place and Policy Institute 2018; 2019), as well as of the Overview of Heritage Lottery Fund Outcomes (Hull 2017 Ltd, 2019). Semi-structured interviews involved senior staff from Hull City Council, Visit Hull and East Yorkshire, the Heritage Lottery Fund (National Lottery Heritage Fund since 2019), Absolutely Cultured Ltd (the post-event legacy company), the University of Hull and local heritage groups.[1] A review of local and national press articles[2] was undertaken to detect positive and negative media narratives about Hull. The guiding themes informing the analysis of interviews and media items for this article were: heritage learning and civic pride; heritage, regeneration and the city's image; maritime heritage; preconceptions and pride; the spectacularisation of heritage; the role of cultural and heritage-related partnerships.[3]

A Northern European port city

Kingston upon Hull is a port city in Northern England (Figures 1 and 2). A 2013 study about medium-sized cities in England portrayed Hull as economically isolated, and characterized by a relatively weak and insufficiently diversified economy (Bolton and Hildreth 2013). However, Hull has been a port city for more than 700 years (Starkey et al. 2017), where trade with Northern Europe was historically the main economic activity,

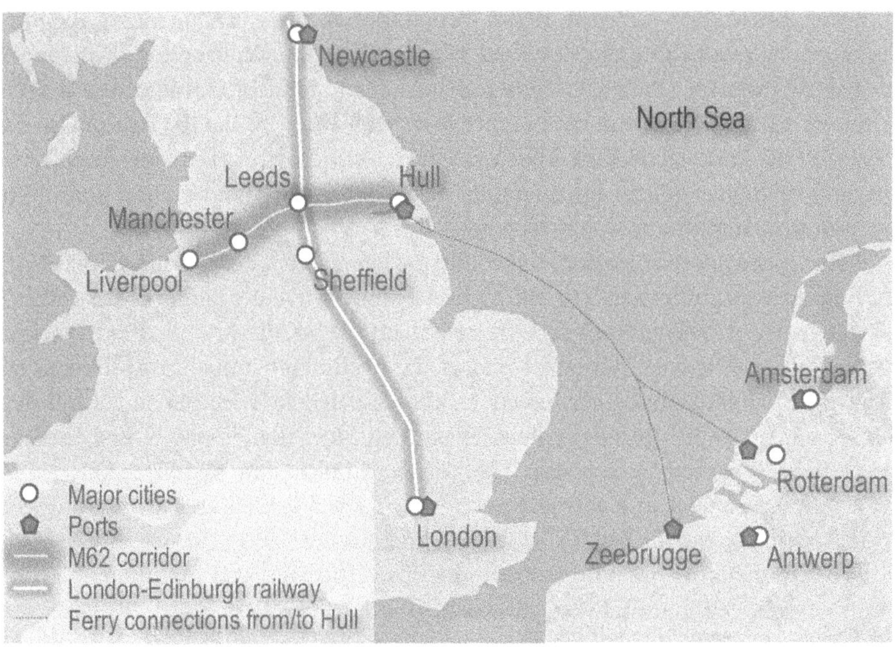

Figure 1. Hull geographical location. Enrico Tommarchi. Background map: Imagery ©2020 TerraMetrics, Map data ©2020.

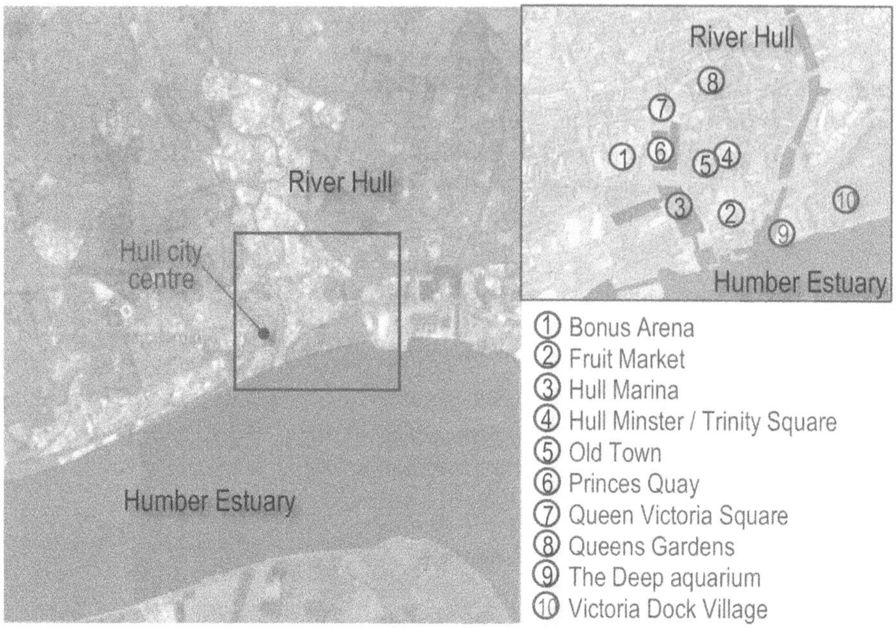

Figure 2. Hull and Hull city centre. Enrico Tommarchi. Background maps: Imagery ©2020 TerraMetrics, Map data ©2020; Imagery ©2020 CNES / Airbus, Getmapping plc, Infoterra Ltd and Bluesky, Landsat / Copernicus, Maxar Technologies, The Geoinformation Group, Map data ©2020.

alongside shipbuilding and whaling. Fishing enjoyed a boom in particular between the Second World War and the 1960s.

However, particularly since the 1970s, a combination of factors including the impact of the oil crisis, overfishing and the Cod Wars with Iceland caused the contraction of Hull's fishing industry (Byrne 2015), which unfolded alongside some of the processes of deindustrialization visible in many other European cities. In the last decades, in comparison with the national average, Hull was characterized by relatively low levels of education, and by higher figures for unemployment, deprivation and mental and physical health problems than the national averages.[4]

Hull can be considered as one of the country's 'left-behind' coastal cities. In 2015, 5 of the 10 English local authorities with the highest proportion of neighbourhoods within the 10% most deprived areas in England were coastal cities (Middlesbrough, Hull, Blackpool, Hartlepool and Liverpool).[5] Since the end of the twentieth century, Hull suffered from negative external perceptions and preconceptions related to unemployment and deprivation – as many other European port cities (Van Hooydonk 2007) – and became 'a city unusually sensitive to its image' (Atkinson, Cooke, and Spooner 2002, 27). In 2003, it was top-ranked in a satirical publication, *The Idler Book of Crap Towns: The 50 Worst Places to live in the UK* (Jordison and Kieran 2003), whose impact on the media gave impetus to prejudices about the city. Hull was also mentioned in media accounts of deprived communities and rustbelt cities (The Economist 2013; Lehmann 2016). Economic decline and deprivation were arguably among the key factors behind the local results of the European Union (EU) referendum in June 2016, where 67.6% of Hull residents taking part voted to leave the EU, compared with 51.9% nationally.

Media comments about Hull's vote in the EU referendum underlined the contrast between Hull's simultaneously inward- and outward-looking attitudes (Clavane 2017) and the recent £310 million investment by German-based company Siemens in a wind turbine manufacturing plant (Conn 2017).

Nevertheless, Hull is still an important port city, where maritime activities include trade and food handling. The Green Port initiative – launched in 2010 by Hull City Council, East Riding of Yorkshire Council, Associated British Ports and partner organizations including the University of Hull – aims at promoting the port of Hull as a hub for the development of renewable energies. The 2013 City Plan envisioned the future of Hull as an energy port city and included the very ambitious goal of the city becoming a world-class visitor destination.

Since the 1980s, Hull City Council has pursued urban regeneration across the city centre and waterfront. Junction Dock and Railway Dock were converted into a marina in 1983 (Figure 3). A shopping centre (Princes Quay) was built on Princes Dock in 1991 (Figure 4). Between the late 1980s and the early 2000s, Victoria Dock was redeveloped into a housing estate. The regeneration of the city centre was prioritized in subsequent policies, in order to strengthen the area's retail offer (Hull City Council 2000; Hull Citybuild 2003; Hull City Council 2008).

The UK City of Culture 2017

Hull City Council bid, unsuccessfully, for the UK City of Culture 2013 and then, this time successfully, for the UK City of Culture 2017. The UK City of Culture (UKCoC), the UK's quadrennial cultural festival, was launched in 2010 with the aim of allowing other UK

Figure 3. Hull Marina was used in 2017 for the Freedom Festival. The Bonus Arena (in the background) was intended as a legacy project of the UKCoC 2017. Enrico Tommarchi.

Figure 4. Princes Quay, with the Maritime Museum in the background. In 2017, technological installations were displayed in the area as part of the Where Do We Go From Here programme. Enrico Tommarchi.

cities to access the potential socio-economic benefits of cultural mega-events that were witnessed in Glasgow, European City of Culture 1990, and Liverpool, European Capital of Culture 2008 (DCMS 2009). Glasgow 1990 was a 'watershed' in the history of the European City/Capital of Culture (ECoC; Bianchini, Albano, and Bollo 2013) as it was the first ECoC delivering a 365-day programme and displaying a focus on urban regeneration (García 2004, 2005; Mooney 2004). Liverpool ECoC 2008 also focused on culture-led urban regeneration (Boland 2010; García, Melville, and Cox 2010; Cox and O'Brien 2012; Connolly 2013). Derry-Londonderry, in Northern Ireland, was the first ever UKCoC in 2013. Hull held the title in 2017, while Coventry will be the next UKCoC in 2021.

The UKCoC is intended, as the ECoC, as a 365-day cultural festival. However, in line with British ECoCs, it has a strong emphasis on urban, economic and/or social regeneration. For example, bidders are asked to describe the 'step changes' (DCMS n.d.) that the event would trigger for the city and the local economy.

UKCoC budgets are on average slightly lower than those of recent ECoCs. The operational budgets of Derry-Londonderry 2013 and Hull 2017 were respectively £18.3 million (€21.53 million, ca €199 per resident, at 2013 exchange rates; Derry City and Strabane District Council 2014) and £32.8 million (€37.27 million, ca €143 per resident, at 2017 exchange rates; Culture, Place and Policy Institute 2018). By comparison, the average budget of the ECoC in the period 2005–2011 was €49 million (nearly €200 per resident; Bianchini, Albano, and Bollo 2013). More recent ECoCs included Donostia-San Sebastián 2016 (€49.6 million, ca €275 per resident), Pafos 2017 (€8.5 m, ca €242 per resident), Aarhus 2017 (€61.9 million, ca €187 per resident), and Leeuwarden 2018

(€104.6 million, ca €870 per resident).[6] ECoC programmes tend to be more international in nature and slightly more effective in attracting international audiences, which accounted for about 1% of total audiences in the case of Hull 2017 (Culture, Place and Policy Institute 2018), in comparison with 3% for Aarhus ECoC 2017 (Degn et al. 2018) and 7% for Leeuwarden ECoC 2018 (EFECTIV Consortium 2019).

The factors which led to Hull securing the title included the extensive participatory and consultation programme and the compelling case that was made in the bid of 'a city coming out of the shadows' that 'needed' the event (Hull UK City of Culture 2017, 2013, 4–5). The bidding process raised initial scepticism (academic1, interview, April 2019) around the idea that Hull could be a cultural capital and could deliver the event. The excitement generated by the designation turned in the planning stage into a mixture of positivity and scepticism (academic4, interview, July 2019), the latter with regard particularly to consultation processes and the implementation of public realm works in the city centre (with Hull being dubbed in the media the 'City of Orange Barriers'). The event was delivered by Hull 2017 Ltd, a separate company established by the City Council.

The Preliminary Outcomes Evaluation Report of Hull UK City of Culture 2017 (Culture, Place and Policy Institute 2018) presents the immediate impacts: more than 2800 events were held across more than 300 venues; there were 5.3 million audience visits; tourist visits in 2017 were estimated to be 6.2 million, against a baseline of 5.5 million in 2016 and a figure of 6.25 million in 2018.[7]

The report observes that 75% of residents said that they were proud to live in Hull, while 71% of them would speak positively about the city. 3 in 4 visitors said that the UKCoC 2017 changed their perceptions of Hull for the better and 61% of them said that they were likely to return. In terms of external perceptions across the UK, the report suggests that 46% of respondents to a national survey said that their perception of the city had changed for the better. However, the report also highlights the fact that the proportion of respondents perceiving Hull as an appealing place to visit increased by a mere 1% (from 50% to 51%) since 2016.

Positive accounts of Hull were about changing internal and external perceptions of the city, increasing visitor numbers and the richness of the Hull 2017 cultural programme (Barnett 2017; BBC 2017a; BBC 2017b; Wonfor 2017). The 2016 *Rough Guide* included Hull among the 10 best cities to visit in the world. As a result of the partnership with the BBC, Hull was also displayed on the national weather map. Press articles about the event show occasional references to the opportunity for Hull to overcome long-standing stigmatization (English 2014; Moore 2016).

Mega-events, heritage and stigmatization: the case of Hull

History and heritage played a pivotal role in Hull's UKCoC 2017 programme. Of the 465 new commissions, 1 in 2 was inspired by history and heritage (Culture, Place and Policy Institute 2018). This central role was not initially envisioned in the 2013 bid (academic2, interview, April 2019). A key factor in this shift was the partnership, established in early 2016, with the Heritage Lottery Fund (HLF), a funding body which in 2019 became the National Lottery Heritage Fund (NLHF). HLF contributed £3 million to the funding of the cultural programme, while its officers were involved in discussions with Hull 2017

Ltd about the role of heritage as a source of inspiration (HLF officer, interview, April 2019).

Local history and heritage were a means to challenge internal and external perceptions of Hull. A key aim was to increase awareness and appreciation of Hull's history and heritage among both residents and visitors (event team member, interview, July 2019). This was achieved through the use of visual arts and theatre projects in particular (Hull 2017 Ltd, 2019). For instance, *Look Up* involved a number of art installations in heritage spaces, such as Trinity Market or Queen Victoria Square. Residents and visitors were encouraged to 'look up' at heritage buildings, which were mobilized 'as a canvas' (tourism board officer, interview May 2019), and to experience them from a fresh perspective. Heritage was also mobilized to address social challenges, as a means to encourage engagement and participation and to boost civic pride (academic2, interview, April 2019; heritage group member, interview, March 2019), for example by featuring in cultural programmes in the city's schools (e.g. *No Limits*).

For many interviewees the fact that the event connected with local stories and narratives made residents feel enabled and motivated to attend and take part in cultural activities, and to engage with the city's heritage (academic1, interview April 2019; HLF officer, interview, April 2019; academic2, interview April 2019; event team member, interview, July 2019; NLHF officer, interview, May 2019). Postcode data about 2017 audiences suggest that the UKCoC was successful in engaging residents from deprived neighbourhoods, except for those parts of the city included in the 10% most deprived areas of the country, and that happiness and life satisfaction ratings rose between 2015 and the beginning of 2017 (Culture, Place and Policy Institute 2018). Some interviewees, however, felt that many heritage-inspired projects tended to engage with well-known historical events and to lever on stereotypes (academic1, interview, April 2019; academic3, interview, August 2019; academic4, interview, July 2019).

Learning and appreciation of heritage

The 2017 Citywide Residents' survey showed that 65.6% of respondents believed that the UKCoC contributed to increasing their knowledge of the city's history and heritage. The audiences of 2017 events largely believed that using arts and cultural activities to communicate aspects of local history and heritage made these more interesting (91.3%) and easier to understand (91.4%; Culture, Place and Policy Institute 2018). However, the 2018 Citywide Residents' survey (Information by Design 2019) showed a decrease in the proportion of residents who rated their knowledge of local heritage as high. This may be related either to a superficial learning process in 2017 or to a more self-critical approach by residents in the light of increased awareness (Culture, Place and Policy Institute 2019).

Urban regeneration and heritage

Urban regeneration and tourism development were viewed by Hull policy makers as a means to counter stigmatization. The UKCoC 2017 acted as a catalyst for external regeneration projects. The Bonus Arena, a 3500-seat venue, was built near the Marina. The Fruit Market (a former wholesale fruit and vegetable market consisting of several heritage buildings located between the city centre and the waterfront) is being redeveloped into a mixed-used urban area and a cultural hub, including cultural facilities such as Humber Street Gallery and Stage@The Dock.

Part of the city's built heritage was showcased through the refurbishment of public space. Public realm improvements in 2015 and 2016 focused on a range of public and heritage spaces in the city centre, such as Queen Victoria Square, Humber Street and Trinity Square (Figures 5, 6 and 7), with the aim of getting the city ready for 2017. These refurbishments helped reconnect heritage assets and made some of them more visible, as in the case of the foundations of the medieval Beverley Gate.

Heritage buildings hosting flagship cultural institutions such as the Ferens Art Gallery and Hull New Theatre also underwent refurbishment and contributed to shaping positive accounts about the city in the media.

The role of maritime heritage

As acknowledged in the city's Cultural Strategy 2016–2026, despite the fact that local policy makers were beginning to mobilize maritime heritage to rebrand the city, its visibility and interpretation remained problematic (Hull City Council 2016, 21; academic3, interview, August 2019). In addition, the city's fishing past had been largely overlooked in public policy in the late twentieth century (academic2, interview, April 2019). At that time, Hull's fishing past could be interpreted as an example of dissonant heritage (Tunbridge and Ashworth 1996), as it recalled painful collective memories related to the many sailors and fishermen who had died at sea and to the decline of the local fishing industry.

Figure 5. The refurbished Queen Victoria Square, one of the areas where public realm investment was concentrated, was the main venue of We Are Hull, as part of the opening Made in Hull programme. The Maritime Museum, the City Hall and the Ferens Art Gallery, key venues in 2017, are located on the square. Enrico Tommarchi.

Figure 6. Humber Street in the Fruit Market area. A partnership between the City Council and property developer Wykeland Beal made it possible to transform an almost completely abandoned area into a cultural hub. Enrico Tommarchi.

Figure 7. Hull Minster and Trinity Square were used as venues for art installations and events in 2017 and beyond (e.g. Hall for Hull, Urban Legends: Northern Lights). Public realm improvements included reflective pools which increased the visibility of historic buildings. Enrico Tommarchi.

Nevertheless, since the end of the twentieth century and in particular in 2017, fishing heritage became prominent, overshadowing other aspects of port activity (academic2, interview, April 2019; academic4, interview, July 2019; City Council officer, interview, April 2019). Some projects and cultural events related to fishing had a positive response from residents. One example is the restoration of the Dead Bod, a 1960s sketch of a dead bird painted by trawlerman Len 'Pongo' Rood and his engineer Gordon Mason on an iron shed at Alexandra Dock. The painting risked disappearing due to the construction of Siemens' wind turbine manufacturing plant. A partnership involving Associated British Ports allowed to save the painting, which was restored and put on public display at Humber Street Gallery, where it was seen by approximately 120,000 people in 2017 (Hull 2017 Ltd, 2019). Another example was *The Last Testament of Lillian Bilocca*, a flagship play performed in the Guildhall as part of the City of Culture programme, which explored activism by a group of local women (led by Lillian Bilocca) against the unsafe and harsh working conditions of Hull's fishermen and sailors in the late 1960s.

Spectacularising heritage

Many cultural events in Hull generated positive media narratives about the city. For example, *Sea of Hull* in 2016 contributed to showcasing some of Hull's heritage spaces and had considerable visibility in the media internationally (BBC 2016; Perraudin 2016): it 'reached 720 million people in 54 countries putting Hull on the map from Australia, Canada, USA, Mexico and Iceland to Finland, China and the Far East, India and most countries across Europe' (ITV 2016). *Sea of Hull* was profiled also during the cultural programme of the UKCoC 2017 through the exhibition *SKIN* at the newly-refurbished Ferens Art Gallery, where pictures of the event were displayed (BBC 2017e). At the start of the City of Culture year, *Made in Hull* involved a spectacular firework display and light shows and projections on heritage buildings. 342,000 people attended the event (Culture, Place and Policy Institute 2018). Also, in January 2017, the *Blade* installation consisted in a 75 m wind turbine rotor manufactured at Hull's Alexandra Dock being displayed in Queen Victoria Square. It was estimated that the installation was seen by 1.1 million people, 420,000 of whom interacted with it (Hull 2017 Ltd, 2019). Later in the year, *Poppies: Weeping Window* was an installation consisting of a cascade of ceramic poppies on the façade of the Maritime Museum, as part of *14–18 NOW*, a national programme of art commissions to mark the centenary of the First World War. This heritage-inspired installation contributed to growing media interest in Hull and the UK City of Culture (Parveen 2017; ITV 2017b).

However, this spectacularisation of heritage should be further problematized, by considering its broader impacts. In the case of *Made in Hull*, the spectacularisation of heritage was nonetheless linked to local stories (academic2, interview, April 2019) – for example, aspects of the event explored painful memories related to the city's fishing history. This generated an emotional response from residents (HLF officer, interview, April 2019), which made it easier to connect with the local population. The success in terms of audience numbers and the visual impact of events involving the spectacularisation of heritage – *Made in Hull* in particular – played a key role in dissolving the scepticism which had been present before 2017 about the ability of local actors to effectively deliver the UKCoC. These spectacular, highly engaging, outdoor and free of charge cultural events also contributed to the 'humanisation of public space' (Culture, Place and Policy Institute 2018, 160). After the widespread refurbishments undertaken within a compressed timescale in 2015 and 2016, such events helped residents reframe in positive terms their relationships with heritage spaces.

Heritage-related partnerships

The role of heritage within the UKCoC programme impacted on external perceptions also in relation to heritage-related partnerships. The work with HLF and Historic England had two main impacts. On the one hand, such partnerships contributed to showcasing the city's heritage – including assets not immediately thought of as 'heritage' (HLF officer, interview, April 2019) – and to generating positive narratives about the city. For example, national newspapers reported that nine new heritage assets were listed during the City of Culture year (BBC 2017d; Halliday 2017), including the Humber Bridge (1981) and the Tidal Surge Barrier (1980). The Old Town was also included in the national Heritage Action Zone programme (ITV 2017a) for economic revitalization and growth in heritage-rich areas. On the other hand, these partnerships meant that local institutions and organizations established closer relationships with national organizations and that the city acquired a higher-profile track record.

Hull as a heritage (port) city? The legacy of 2017

Heritage-inspired cultural events along the lines of those taking place during the City of Culture year continued in 2018 and 2019, delivered by the legacy company Absolutely Cultured Ltd. *Dominoes* consisted of a route of large concrete white domino pieces stretching across some of the city centre's iconic locations (Preston and Winter 2018). Similarly, *Urban Legends: Northern Lights*, mobilized public and heritage spaces and less well-known areas of the city, where light shows celebrated Hull's relationships with Northern Europe and Scandinavia (Young 2018). These events generated a good response from the public, with audience figures of over 20,000 for the one-day event *Dominoes* and over 50,000 for *Urban Legends: Northern Lights* (Culture, Place and Policy Institute 2019). In early 2018, Hull enjoyed a certain extent of national media coverage also because of a graffiti reportedly by street artist Banksy (e.g. BBC 2018). In 2019, the local intangible heritage of myths, stories and childhood memories was spectacularised by Absolutely Cultured Ltd through *The Witching Hour*, which was also staged in Wirral, the Liverpool City Region's 2019 Borough of Culture (Davis 2019).

The case of Hull is also an example of the role of heritage in fostering participation and engagement in cultural activities (Ponzini et al. 2019). These events created arts-heritage fusions, hybrids or 'third spaces' (Froggett, Muller, and Bennett 2019), which encouraged residents and visitors to engage with less well-known heritage assets and spaces (Tommarchi and Bianchini 2020) and thus to appreciate a broader range of aspects of the city's history.

Maritime heritage and stigmatization

The mobilization of Hull's maritime heritage in relation to internal and external perceptions of the city is set to be further explored through *Hull: Yorkshire's Maritime City*, a heritage-inspired project perceived as another potential game changer (Campbell 2019). The £27.4 million scheme is led by Hull Culture and Leisure (HCAL) and Hull City Council. The National Lottery Heritage Fund (NLHF) is contributing with £13.6 million to the project. It will restore a number of maritime heritage assets, namely the Maritime Museum, the adjacent Dock Office Chambers, the North End Shipyard, the 1960s trawler Arctic Corsair and the Spurn Lightship. The project aims at creating a coherent, maritime-related cultural offer, to broaden perceptions of the city's heritage beyond stereotypes and established interpretations and to attract visitors and cruise passengers.[8] In addition, it represents the opportunity to 'rebalance' maritime heritage narratives (City Council officer, interview, April 2019). *Hull: Yorkshire's Maritime City* is also seen by policy makers as a means to generate positive media narratives about the city (BBC 2017c; BBC 2019), in relation to cultural attractiveness, urban regeneration and the delivery of an ambitious project.

This scheme shows how maritime heritage has been shifting from being an element of dissonant heritage linked with stigmatization – which was initially chosen not to be explored in 2017 – to a cultural resource linked with civic pride, leisure and tourism. However, with regard to its connection with tourism and economic regeneration, the project may potentially raise future issues of authenticity (Jones and Ponzini 2018; Ponzini et al. 2019), concerning the ways in which Hull's maritime history and heritage will be explored and mobilized.

Hull's image in the medium term

The longer-term impacts of the UK City of Culture 2017 need to be problematized. Happiness and life satisfaction indicators rose in the run up to the event, but they returned to pre-2016 levels by the end of the City of Culture year (Culture, Place and Policy Institute 2018). In 2018, 71% of residents said that they were proud to live in Hull, against a figure of 75% in 2017, nonetheless accounting for a 1% increase since 2016 (Culture, Place and Policy Institute 2019). Despite the concept of the 'city as a venue', which resulted in more than 300 cultural venues in 2017, cultural events tended to be concentrated in the city centre. Understandably, this is a recurrent feature of culture-led urban regeneration and it appears in line with Hull City Council's long-standing strategy of prioritizing the regeneration of the city centre. However, such spatial concentration of cultural activities and investment risks hijacking resources to the detriment of cultural activities and heritage conservation in other parts of the city, as well as opening a reputational gap between the city centre and the periphery.

Some interviewees suggested that, after the event, scepticism among the local population has grown once again, perhaps fuelled by factors that are unrelated to the event itself and not Hull-specific. Due to the local results of the 2016 EU referendum and the upsurge of the Brexit Party in the 2019 European elections, some accounts of Hull as a city which tends to be inward-looking, pro-Brexit, unwelcoming to immigrants and declining – with little trace of City of Culture optimism – began to emerge in the media (e.g. Araujo 2017; Charles 2017; Clavane 2017; Stephenson 2020). This risks undermining the positive impact on external perceptions of the 2017 UKCoC and of media narratives about the event and its outcomes. The national crisis of retail (Butler 2019; Simpson 2019), substantially accelerated by the shift to online shopping related to the COVID-19 pandemic, is not specifically associated with Hull. However, shop closures in the city centre translate into a growing number of unused buildings, including historic ones. This process risks producing fragmentation across the townscape, reversing the regenerative process fuelled by the event, as well as the outcomes of heritage-inspired events and policies aiming at reconnecting urban heritage assets (academic2, interview, April 2019). These negative external factors may have an adverse effect on internal and external perceptions of the city.

Conclusions

This paper has examined a cultural mega-event, Hull UK City of Culture 2017, in a heritage-rich city suffering from long-standing stigmatization due to structural socio-economic challenges emerged in the closing decades of the twentieth century. The article aims to show heritage was mobilized within the event and contributed to changing internal and external perceptions of the city.

Hull is an example of how policy makers and event organizers can shape cultural mega-events to address territorial stigmatization, for example by circulating positive narratives about the city across the media and through word of mouth, by reframing aspects of dissonant heritage, and by forming new local or national partnerships to invest in new heritage-led schemes.

The experience of Hull 2017 shows that when arts and cultural activities engage with heritage in creative ways, learning and participation can impact on internal and external perceptions of a city through increased awareness and appreciation of local history and heritage itself. In addition, the reconnection of heritage assets across the townscape had revitalizing effects in Hull, with positive impacts on perceptions of the city. The spectacularisation of heritage had more nuanced outcomes beside the mere generation of positive media contents, for example in relation to awareness of and engagement with heritage, challenging scepticism, and humanizing refurbished public spaces. Finally, heritage-related partnerships with external or national organizations may have longer-term positive impacts on the reputation of local cultural organizations, their level of ambition and their ability to attract funding.

There are policy implications and research perspectives arising from this case study. One key aspect is the link between the mobilization of maritime heritage to counter stigmatization and the risks of cultural standardization of a port city, where certain aspects of local maritime cultures are neglected or overridden (Tommarchi 2019). Another issue is the potential conflict between the imaginaries of outward-looking cities of culture – promoted through cultural mega-events – and media accounts of inward-looking places due to the rise of right-wing populist and nationalistic political parties locally. In the case of port cities like Hull, there is a contradiction between the allegedly independent and rebellious spirit associated with such places (Van Hooydonk 2007; Starkey et al. 2017) and emerging anti-immigrant and anti-cosmopolitan political cultures (Bianchini and Tommarchi 2020). The politicization and mediatization of cultural mega-events and local heritage in the current political climate (Ponzini et al. 2019) – also in relation to stigmatization and socio-economic regeneration – therefore need further exploration, in particular by considering contemporary examples of 'invention of traditions' (Hobsbawm and Ranger 1994) at city level.

Finally, an open question concerns the legacy of heritage-inspired cultural mega-events in terms of perceptions of the city. The case of Hull shows how cultural mega-events on their own are not enough to eradicate the prejudices and preconceptions which lie behind stigmatization. Without a clear, long-term strategy from the outset, momentum can be lost quickly, and the positive outcomes of the cultural mega-event can soon evaporate. Stigmatization may arise rapidly once again, perhaps fuelled by unrelated and unpredictable external factors.

Notes

1. Ten semi-structured interviews were undertaken in March-August 2019 to prepare a case study about heritage opportunities and threats in relation to Hull 2017 as part of the HOMEE project. The role of heritage in countering negative internal and external perceptions of the city was a key theme emerging from the interviews. This research strand about on heritage, territorial stigmatization and cultural mega events benefitted from the authors' work on the evaluation of Hull 2017 (undertaken by the Culture, Place and Policy Institute of the University of Hull).
2. The media analysis focused on online items from local and national sources issued in 2014–2019. Articles published by the local newspaper (*Hull Daily Mail*) were analysed alongside others from local newspapers from across Northern England mentioning Hull 2017. The analysis of the national media encompassed national newspapers (*The Guardian, The Independent*), broadcasters (BBC, ITV), and political magazines (*New Statesman, The Economist, The New European*). Critical discourse analysis (Wodak 2001) was used to detect recurrent discourses and storylines (Hajer 2003).
3. During the research process, the UK City of Culture 2017 has been perceived as relatively less important for Hull since 2018, mostly due to unrelated factors such as Brexit, the impact of austerity policies and lately the COVID-19 pandemic. Conversely, heritage-led regeneration has become increasingly important, as demonstrated by the interest around the *Hull: Yorkshire's Maritime City* project.
4. In 2019, 27.2% of the local population held a NVQ4 qualification or above, while 8.9% had no qualification, in comparison with 40.3% and 7.7% in Great Britain; the unemployment rate in 2019 was 6.6% against 3.9% nationally (Source: Office for National Statistics). Hull displayed the third highest proportion of neighbourhoods among the 10% most deprived in England (Source: Department for Communities and Local Government, 2015). The proportion of adults diagnosed with chronic mental health problems is above the national average (source: Hull Data Observatory, PHOF and DWP), as are figures for stroke, coronary heart disease and respiratory problems (source: Hull City Council, 2018).
5. Source: Department for Communities and Local Government, 2015.
6. As noted by Bianchini, Albano, and Bollo (2013) for the 2005–2011 period, ECoCs in the last twenty years have included small and medium-sized cities. Final operational budget figures are from Fox and Rampton (2017), Directorate-General for Education, Youth, Sport and Culture (2018) and EFECTIV Consortium (2019).
7. Source: Visit Hull and East Yorkshire (2019).
8. A cruise terminal is planned to be built near the iconic aquarium The Deep, at walking distance from Hull's Old Town.

Acknowledgments

Sections of this paper are based on and supported by the research conducted within the JPI Cultural Heritage research project 'HOMEE – Heritage Opportunities/ Threats within Mega-Events in Europe', with a grant issued by the UK's Arts and Humanities Research Council.

Disclosure statement

No potential conflict of interest was reported by the author(s).

References

Araujo, F. 2017. "Hull Revisited: What Happens When a Brexit Stronghold Becomes City of Culture?" *New Statesman*. https://www.newstatesman.com/politics/uk/2017/01/hull-revisited-what-happens-when-brexit-stronghold-becomes-city-culture.

Atkinson, D., S. Cooke, and D. Spooner. 2002. "Tales from the Riverbank: Place-Marketing and Maritime Heritages." *International Journal of Heritage Studies* 8 (1): 25–40. doi:10.1080/13527250220119910.

Barnett, D. 2017. "Kingston upon Hull: Culture Can Be a Catalyst for Real Change in any City." *The Independent*. https://www.independent.co.uk/arts-entertainment/culture-can-be-a-catalyst-for-real-change-in-any-city-a7541656.html.

BBC. 2016. "Thousands Strip for Sea of Hull Mass Nude Photograph." *BBC News*. https://www.hulldailymail.co.uk/whats-on/whats-on-news/gallery/sea-hull-pictures-21-best-70440.

BBC. 2017a. "Hull's 2018 UK City of Culture Year 'Life-Enhancing'." *BBC News*. https://www.bbc.co.uk/news/uk-england-humber-42501917.

BBC. 2017b. "Hull's City of Culture Events 'Raise Bar for Successor'." *BBC News*. https://www.bbc.co.uk/news/uk-england-humber-40420313.

BBC. 2017c. "Hull's Maritime Attractions Given £27.5 m for Renovation Work." *BBC News*. https://www.bbc.co.uk/news/uk-england-humber-40408516.

BBC. 2017d. "Humber Bridge among Nine Hull Locations to Be Given Listed Status." *BBC News*. https://www.bbc.co.uk/news/uk-england-humber-40623380.

BBC. 2017e. "Sea of Hull Nudes to Be Flagship City of Culture Exhibition." *BBC News*. https://www.bbc.co.uk/news/uk-england-humber-39052795.

BBC. 2018. "Latest Banksy Graffiti Unveiled in Hull." *BBC News*. https://www.bbc.co.uk/news/uk-england-humber-42841320.

BBC. 2019. "Images of Hull's Maritime Attractions Revamp Plans Unveiled." *BBC News*. https://www.bbc.co.uk/news/uk-england-humber-48042095.

Bélanger, A. 2000. "Sport Venues and the Spectacularization of Urban Spaces in North America. The Case of the Molson Centre in Montreal." *International Review for the Sociology of Sport* 35 (3): 378–397. doi:10.1177/101269000035003009.

Bianchini, F., R. Albano, and A. Bollo. 2013. "The Regenerative Impacts of the European City/Capital of Culture Events." In *The Routledge Companion to Urban Regeneration*, edited by M. Leary, and J. McCarthy, 515–525. London: Routledge.

Bianchini, F., and E. Tommarchi. 2020. "Urban Activism and the Rethinking of Cities' Cultural Policies in Europe." In *Kann Kultur Politik? – Kann Politik Kultur? Warum wir wieder mehr über Kulturpolitik sprechen sollten*, edited by M. Wimmer, 156–166. Vienna: De Gruyter.

Boland, P. 2010. "'Capital of Culture—You Must Be Having a Laugh!' Challenging the Official Rhetoric of Liverpool as the 2008 European Cultural Capital." *Social and Cultural Geography* 11 (7): 627–645. doi:10.1080/14649365.2010.508562.

Boland, P., B. Murtagh, and P. Shirlow. 2019. "Fashioning a City of Culture: 'Life and Place Changing' or '12 Month Party'?" *International Journal of Cultural Policy* 25 (2): 246–265. doi:10.1080/10286632.2016.1231181.

Bolton, T., and P. Hildreth. 2013. *Mid-Sized Cities: Their Role in England's Economy*. London: Centre for Cities.

Bourdieu, P. 1991. *Language and Symbolic Power*. Cambridge: Polity Press.

Butler, S. 2019. "Thousands of UK Shops Left Empty as High Street Crisis Deepens." *The Guardian*. https://www.theguardian.com/business/2019/may/02/thousands-of-uk-shops-left-empty-as-high-street-crisis-deepens.

Byrne, J. 2015. "After the Trawl: Memory and Afterlife in the Wake of Hull's Distant-Water Fishing Industry." *International Journal of Maritime History* 27 (4): 816–822. doi:10.1177/0843871415610281.

Campbell, J. 2019. "Hull Tourism Boost as Game-Changing Maritime History Project Becomes Reality with £13.6 m National Lottery Grant." *Hull Daily Mail.* Retrieved from: https://www.hulldailymail.co.uk/news/hull-east-yorkshire-news/hull-tourism-maritime-national-lottery-3380519.

Charles, A. 2017. "How Hull Went from Crap Town to City of Culture – And What It Says about Brexit Britain." *The Conversation.* http://theconversation.com/how-hull-went-from-crap-town-to-city-of-culture-and-what-it-says-about-brexit-britain-86818.

Chhabra, D., R. Healy, and E. Sills. 2003. "Staged Authenticity and Heritage Tourism." *Annals of Tourism Research* 30 (3): 702–719. doi:10.1016/S0160-7383(03)00044-6.

Clavane, A. 2017. "Brexit Heartland and City of Culture Hull Remains in Dangerous Waters." *The New European.* https://www.theneweuropean.co.uk/top-stories/brexit-heartland-and-city-of-culture-hull-remains-in-dangerous-waters-1-5322162.

Conn, D. 2017. "Hull after Brexit – Will the City of Culture Regret Voting out?" *The Guardian.* https://www.theguardian.com/politics/2017/apr/02/hull-after-brexit-will-the-city-of-culture-regret-voting-out.

Connolly, M. 2013. "The 'Liverpool Model(s)': Cultural Planning, Liverpool and Capital of Culture 2008." *International Journal of Cultural Policy* 19 (2): 162–181. doi:10.1080/10286632.2011.638982.

Cox, T., and D. O'Brien. 2012. "The "Scouse Wedding" and Other Myths: Reflections on the Evolution of a "Liverpool Model" for Culture-Led Urban Regeneration." *Cultural Trends* 21 (2): 93–101. doi:10.1080/09548963.2012.674749.

Culture, Place and Policy Institute. 2018. *Cultural Transformations: The Impacts of Hull UK City of Culture 2017. Preliminary Outcomes Evaluation. March 2018.* Hull: University of Hull.

Culture, Place and Policy Institute. 2019. *Cultural Transformations. The Impacts of Hull UK City of Culture 2017. Main Evaluation Findings and Reflections. November 2019.* Hull: University of Hull.

Davis, L. 2019. "Witching Hour Night Time Spectacular with Pyro Coming to Birkenhead Park." *Liverpool Echo.* https://www.liverpoolecho.co.uk/whats-on/arts-culture-news/witching-hour-night-time-spectacular-16180549.

DCMS. 2009. *City of Culture. Working Group Report. June 2009.* London: DCMS.

DCMS. n.d. *UK City of Culture 2017. Guidance for Bidding Cities.* https://assets.publishing.service.gov.uk/government/uploads/system/uploads/attachment_data/file/89369/UK_City_of_Culture_2017_Guidance_and_Criteria.pdf.

Dean, J., and A. Hastings. 2000. *Challenging Images. Housing Estates, Stigma and Regeneration.* Bristol: The Policy Press.

De Cesari, C., and R. Dimova. 2019. "Heritage, Gentrification, Participation: Remaking Urban Landscapes in the Name of Culture and Historic Preservation." *International Journal of Heritage Studies* 25 (9): 863–869. doi:10.1080/13527258.2018.1512515.

Degn, H., L. Hansen, N. Thomsen, L. Nordentoft, J. Xie, and M. Hansen. 2018. *Aarhus 2017. Before – During – After. A research-based evaluation of the effects of the European Capital of Culture project.* https://projects.au.dk/aarhus2017/research-based-evaluation/main-report/.

Derry City and Strabane District Council. 2014. *Post Project Evaluation of City of Culture 2013.* http://meetings.derrycityandstrabanedistrict.com/documents/s20721/Appendix%202%20Final%20PPE%202018.pdf.

Di Vita, S., and M. Wilson. 2020. *Planning and Managing Smaller Events: Downsizing the Urban Spectacle.* London: Routledge.

Directorate-General for Education, Youth, Sport and Culture. 2018. *European Capital of Culture – 2017 Ex-post Evaluation.* https://ec.europa.eu/programmes/creative-europe/sites/creative-europe/files/ecoc_2017_evaluation_-_executive_summary_enwithlo.pdf.

EFECTIV Consortium. 2019. *Ex-Post Evaluation of the 2018 European Capitals of Culture – Executive Summary.* https://op.europa.eu/en/publication-detail/-/publication/93aa14ef-1b67-11ea-8c1f-01aa75ed71a1/language-en.
Eisinger, P. 2000. "The Politics of Bread and Circuses: Building the City for the Visitor Class." *Urban Affairs Review* 35 (3): 316–353. doi:10.1177/107808740003500302.
English, C. 2014. "To Hull and Back: The Rebirth of Britain's Poorest City." *The Guardian.* https://www.theguardian.com/cities/2014/sep/11/-sp-to-hull-and-back-the-rebirth-of-britains-poorest-city.
Fox, T., and J. Rampton. 2017. *Ex-post Evaluation of the 2016 European Capitals of Culture. Final report.* https://ec.europa.eu/programmes/creative-europe/sites/creative-europe/files/files/ecoc-2016-evaluation-en_0.pdf.
Froggett, L., L. Muller, and J. Bennett. 2019. "The Work of the Audience: Visual Matrix Methodology in Museums." *Cultural Trends* 28 (2-3): 162–176. doi:10.1080/09548963.2019.1617939.
García, B. 2004. "Cultural Policy and Urban Regeneration in Western European Cities: Lessons from Experience, Prospects for the Future." *Local Economy* 19 (4): 312–326. doi:10.1080/0269094042000286828.
García, B. 2005. "Deconstructing the City of Culture: The Long-Term Cultural Legacies of Glasgow 1990." *Urban Studies* 42 (5-6): 841–868. doi:10.1080/00420980500107532.
García, B., R. Melville, and T. Cox. 2010. *Creating an Impact: Liverpool's Experience as European Capital of Culture.* Liverpool: Impact 08, University of Liverpool and Liverpool John Moores University.
Goffman, E. 1963. *Stigma: Notes on the Management of Spoiled Identity.* New York: Simon and Schuster.
Hajer, M. 2003. "A Frame in the Fields: Policymaking and the Reinvention of Politics." In *Deliberative Policy Analysis. Understanding Governance in the Network Society*, edited by M. Hajer, and H. Wagenaar, 88–110. Cambridge: Cambridge University Press.
Halliday, J. 2017. "Hull Landmarks to Get Listed Status in Boost to City of Culture." *The Guardian.* https://www.theguardian.com/uk-news/2017/jul/17/hull-landmarks-to-be-listed-boost-city-of-culture-status.
Hewison, R. 1987. *The Heritage Industry: Britain in a Climate of Decline.* London: Methuen.
Hobsbawm, E., and T. Ranger. 1994. *The Invention of Tradition.* Cambridge: Cambridge University Press.
Hull 2017 Ltd. 2019. *Evaluation Report. Overview of Heritage Lottery Fund Outcomes. Hull UK City of Culture 2017.* Hull: Absolutely Cultured Ltd.
Hull City Council. 2000. *Hull Cityplan: Adopted May 2000.* Hull: Development Plans Team Transportation and Planning Division Regeneration and Development Directorate.
Hull City Council. 2008. *City Centre Area Action Plan. Incorporating Citywide Policies. Submission Document. July 2008.* Hull: Hull City Council.
Hull City Council. 2014. *Hull City Centre Public Realm Strategy.* Hull: Hull City Council.
Hull City Council. 2016. *Cultural Strategy 2016-2026.* Hull: Hull City Council.
Hull City Council. n.d. *Hull City Plan.* http://cityplanhull.co.uk/.
Hull Citybuild. 2003. *The Renaissance of Hull City Centre.* Hull: Hull City Council.
Hull UK City of Culture 2017. 2013. *Tell the World. Final bid.* Hull: Hull City Council.
Information by Design. 2019. *Residents Survey 2018.* Hull: Information by Design.
ITV. 2016. "Sea of Hull Continues to Generate Exposure for Hull." *ITV News.* https://www.itv.com/news/calendar/2016-07-21/sea-of-hull-continues-to-generate-exposure-for-hull.
ITV. 2017a. "Heritage Status for Hull's Old Town and Industrial South Yorkshire Village Elsecar." *ITV News.* https://www.itv.com/news/calendar/2017-03-20/heritage-status-for-hulls-old-town-and-industrial-south-yorkshire-village-elsecar/.
ITV. 2017b. "Over 300,00 Attend Hull City of Culture Opening." *ITV News.* https://www.itv.com/news/calendar/2017-01-08/over-300-00-attend-hull-city-of-culture-opening/.
Jones, Z. 2020. *Cultural Mega-Events: Opportunities and Risks for Heritage Cities.* London: Routledge.

Jones, Z., and D. Ponzini. 2018. "Mega-events and the Preservation of Urban Heritage: Literature Gaps, Potential Overlaps, and a Call for Further Research." *Journal of Planning Literature* 33 (4): 433–450. doi:10.1177/0885412218779603.

Jordison, S., and D. Kieran. 2003. *The Idler Book of Crap Towns: The 50 Worst Places to Live in the UK.* London: Boxtree.

Kallin, H., and T. Slater. 2014. "Activating Territorial Stigma: Gentrifying Marginality on Edinburgh's Periphery." *Environment and Planning A: Economy and Space* 46: 1351–1368. doi:10.1068/a45634.

Kirkness, P., and A. Tijé-Dra. 2017. "Conclusion. Tainted Urban Spaces at the Intersection of Urban Planning, Politics of Identity and Urban Capitalism." In *Negative Neighbourhood Reputation and Place Attachment. The Production and Contestation of Territorial Stigma*, edited by P. Kirkness, and A. Tijé-Dra, 252–256. London: Routledge.

Lehmann, A. 2016. "The Rise of Trump Comes as No Surprise on England's Disaffected East Coast." *The Guardian.* https://www.theguardian.com/uk-news/2016/nov/12/trumps-rise-no-surprise-englands-disaffected-east-coast-hull-grimsby.

Ley, D. 1983. *A Social Geography of the City.* New York: Harper and Row.

Mooney, G. 2004. "Cultural Policy as Urban Transformation? Critical Reflections on Glasgow, European City of Culture 1990." *Local Economy* 19 (4): 327–340. doi:10.1080/0269094042000286837.

Moore, R. 2016. "City of Culture 2017: Will Hull Have the Last Laugh?" *The Guardian.* https://www.theguardian.com/culture/2016/feb/14/hull-uk-city-of-culture-2017-rowan-moore.

Nedélec, P. 2017. "The Stigmatisation of Las Vegas and Its Inhabitants. The Other Side of the Coin." In *Negative Neighbourhood Reputation and Place Attachment. The Production and Contestation of Territorial Stigma*, edited by P. Kirkness, and A. Tijé-Dra, 9–26. London: Routledge.

Parveen, N. 2017. "Fireworks Launch Hull's Year as the UK's City of Culture." *The Guardian.* https://www.theguardian.com/uk-news/2017/jan/01/fireworks-launch-hull-uk-city-of-culture.

Paton, K. 2018. "Beyond Legacy: Backstage Stigmatisation and 'Trickle-Up' Politics of Urban Regeneration." *The Sociological Review* 66 (4): 919–934. doi:10.1177/0038026118777449.

Paton, K., G. Mooney, and K. McKee. 2012. "Class, Citizenship and Regeneration: Glasgow and the Commonwealth Games 2014." *Antipode* 44 (4): 1470–1489. doi:10.1111/j.1467-8330.2011.00966.x.

Perraudin, F. 2016. "Thousands Strip Naked in Hull for Spencer Tunick Photographs." *The Guardian.* https://www.theguardian.com/artanddesign/2016/jul/09/thousands-strip-naked-in-hull-for-spencer-tunick-photographs.

Ponzini, D., Z. Jones, F. Bianchini, E. Tommarchi, J. Georgi-Tzortzi, E. Dova, A. Sivitanidou, et al. 2019. *HOMEE Literature Review of Mega Events Addressing Cultural Heritage Issues.* http://www.tau-lab.polimi.it/homee-literature-review-regarding-mega-events-cultural-heritage/.

Preston, F., and P. Winter. 2018. "Relive Hull's Spectacular Dominoes Fall as Thousands Enjoy a Brilliant Day in Our Great City." *Hull Daily Mail.* https://www.hulldailymail.co.uk/whats-on/whats-on-news/dominoes-hull-timings-route-traffic-1886588.

Quinn, B. 2005. "Arts Festivals and the City." *Urban Studies* 42 (5-6): 927–943. doi:10.1080/00420980500107250.

Richards, G., and J. Wilson. 2004. "The Impact of Cultural Events on City Image: Rotterdam, Cultural Capital of Europe 2001." *Urban Studies* 41 (10): 1931–1951. doi:10.1080/0042098042000256323.

Simpson, E. 2019. "High Street: How Many UK Shops Have Closed?" *BBC News.* https://www.bbc.co.uk/news/business-49349703.

Slater, T., and N. Anderson. 2012. "The Reputational Ghetto: Territorial Stigmatisation in St Paul's, Bristol." *Transactions of the Institute of British Geographers* 37 (4): 530–546. doi:10.1111/j.1475-5661.2011.00490.x

Starkey, D., D. Atkinson, B. McDonagh, S. McKeon, and E. Salter, eds. 2017. *Hull. Culture, History, Place.* Liverpool: University of Liverpool Press.

Stephenson, B. 2020. "There Is Nothing Happening in Hull to Mark Brexit Day on Friday." *Hull Daily Mail.* https://www.hulldailymail.co.uk/news/hull-east-yorkshire-news/brexit-day-celebrations-european-union-3785544.

The Economist. 2013. "The Urban Ghosts." *The Economist.* https://www.economist.com/britain/2013/10/12/the-urban-ghosts.

Tommarchi, E. 2019. "Exploring Port-City Relationships through Event-Led Urban Regeneration." *Planning for Transition. Book of Papers.* 2019 AESOP Conference, Venice 9-13 July 2019, 3423-3434. https://www.aesop2019.eu/2019/12/01/book-of-papers/.

Tommarchi, E., and F. Bianchini. 2020. "Hull UK City of Culture 2017." In *Mega-Events and Heritage: The Experience of Five European Cities,* edited by D. Ponzini, F. Bianchini, J. Georgi-Tzortzi, and J. Sanetra-Szeliga, 184–254. Krakow: International Cultural Centre. www.tau-lab.polimi.it.

Tunbridge, J., and G. Ashworth. 1996. *Dissonant Heritage: The Management of the Past as a Resource in Conflict.* Chichester: John Wiley and Sons.

Van Hooydonk, E. 2007. *Soft Values of Seaports. A Strategy for the Restoration of Public Support of Seaports.* Antwerp: Garant.

Wacquant, L. 2007. "Territorial Stigmatization in the Age of Advanced Marginality." *Thesis Eleven* 91 (1): 66–77. doi:10.1177/0725513607082003.

Wacquant, L. 2008. *Urban Outcasts. A Comparative Sociology of Advanced Marginality.* Cambridge: Polity Press.

Wacquant, L., T. Slater, and V. Pereira. 2014. "Territorial Stigmatization in Action." *Environment and Planning A: Economy and Space* 46: 1270–1280. doi:10.1068/a4606ge.

Walmsley, D. 1988. *Urban Living. The Individual in the City.* Harlow: Longman Scientific and Technical.

Wodak, R. 2001. "What CDA Is About – A Summary of its History, Important Concepts and Its Developments." In *Methods of Critical Discourse Analysis,* edited by R. Wodak, and M. Meyer, 1–13. London: Sage.

Wonfor, S. 2017. "Hull UK City of Culture 2017: What's Worth Travelling for?" *Chronicle Live.* https://www.chroniclelive.co.uk/whats-on/arts-culture-news/hull-uk-city-culture-2017-12415904.

Young, A. 2018. "Urban Legends: Northern Lights to Bring Made in Hull Magic Back with Huge Sound and Light Festival." *Hull Daily Mail.* https://www.hulldailymail.co.uk/news/hull-east-yorkshire-news/urban-legends-northern-lights-hull-2171937.

Zukin, S. 1995. *The Cultures of Cities.* Oxford: Blackwell.

Cultural events and heritage policy for the Milan Expo 2015: experimental intersections between mega-event and city

Stefano Di Vita

ABSTRACT
The article observes the case of the Expo 2015 by focusing on side cultural events and heritage. By critically drawing on mega-event and smaller event literature, it is argued that new experiments beyond the mega-event itself affected Milan's transformation and repositioning. In particular, the contribution provides new evidence about the connection between the uncertain planning process for the Universal Exhibition 2015 and the enhancement of historic places in both the host city and region. Furthermore, it discusses Milan's urban and regional development trends and agendas turning toward leisure tourism and festivalization in this period.

1. Introduction: heritage and side events beyond the mega

The mega-event of the Milan Expo 2015 has been studied from different perspectives, mostly by concentrating on its core activities: on the one hand, the development and legacy of the Expo site, that has been placed in a large and peri-urban exhibition area (around 1 million sqm) near the new Fairground, in the North-Western transect of the city; on the other hand, the related networks, such as transport, green, and digital infrastructures, as well as cultural, economic, and social activities. However, to date, limited attention has been given to the impact and effects that a set of side activities, policies and plans have had on specific heritage sites (mainly in the urban core, but also in the urban region) and more broadly, within a longer-term trend of urban transformation, repositioning, and touristification of Milan.

The Milan Expo 2015 was open for six months (May-October 2015) and focused on the theme 'Feeding the Planet, Energy for Life', also activating the public debate on urban agriculture. The official exhibition was celebrated in a specific new venue, that attracted more visitors than expected (21.5 million instead of 20 million). Besides the Expo site, the event and its location were originally intended as drivers for the redevelopment of public spaces, and the development of new transport infrastructures (metro lines, train lines, motorways) and green infrastructures (water ways, cycle-paths) at the urban and regional scales. Whilst the total expenditure was about 14,780 million euro, mainly public (Expo 2015 Spa 2018), the added value was estimated at 13,900 million euro, the additional production into 31,600 million euro, the new jobs as 115,000 units, and the new firms as

10,000 units (Dell'Acqua et al. 2016). However, despite the fact that the post-event transformation for the reuse of the Expo site faced significant problems during the planning process and is still under development (Di Vita and Morandi 2018), and the relaunch of urban-rural and metropolitan connections failed, the Expo 2015 undeniably accelerated the process of international repositioning of the city and its region as a leisure tourism destination: from the opening of new museums, to the festivalization of the city centre in particular (in the bidding/planning/legacy period, tourism grew from 3.3 million arrivals in 2007, to 5.3 million in 2015 and 5.5 million in 2017; Di Vita and Ponzini 2020).

This paper provides evidence for explaining this process as partly parallel to the official exhibition plans and infrastructures, and connected to the role played by certain heritage sites, which were used for Expo-related, cultural and entertainment activities (as for the pedestrianization of the area in front of the Sforza Castle where the main in-town contact point called Expo Gates was located), or the renewal of the old city docks as part of nightlife and cultural economy (the so-called Darsena); furthermore, a massive cultural-event initiative was developed in several parts of both the municipal and metropolitan areas (sometimes previously unknown) thanks to the online platform of ExpoinCittà, that was launched by the Milan Municipality and Chamber of Commerce with over 46,000 cultural side-events in over 1,000 sites across the urban core and region.

By drawing on existing critical literature regarding mega-events and smaller events and their relationship with the city (section 2), the article casts new light on the aspects of cultural heritage sites as well as of a myriad of side-events that contributed to the transformation of Milan and its agenda before, during and after the 2015 mega-event. It is indeed clear that these wider processes of involvement of economic and cultural players affected the governance expertise and political constituency (for example the new Mayor, elected in 2016, was the Expo CEO and successfully promoted the bid of Milano-Cortina to the 2026 Winter Olympics). While providing a detailed picture of the planning and legacy of the 2015 Expo (section 3), this contribution identifies the reasons for the incremental experimentation with heritage and side events (section 4). At the same time, it critically discusses the positive and negative outcomes and effects within the context of growing crisis and demands for innovation of mega-events due to their environmental and socio-economic unsustainability. In the concluding sections, the article discusses whether cultural and heritage-based side-events can mitigate the negative impact of mega-events and influence the urban agenda and consensus building processes (section 5), which is relevant for the ongoing international debates regarding mega-events.

2. Mega-events, smaller events, and cultural heritage: urban connections and conflicts

The interest in mega-events has spread in policy and research since the 1990s, from advanced economy countries to developing countries, within local processes of urban de-industrialization, and world processes of economic and cultural globalization. For years, the Olympics, Expos, and FIFA World Cup have been considered important catalysts of economic and human resources, as well as drivers of economic and cultural development, urban change projects, and strategic repositioning of host cities and regions in the 'global map'. This phase has been followed by growing skepticism and protests (Lauermann 2019). Problems of gigantism and disruption of mega-events have contributed to the rise of their negative perception and dissatisfaction, that is due to high and unsustainable costs of organization and construction, difficulties in the reuse of dedicated and often over-sized facilities and infrastructures, diversion of public support to ordinary local services, increase of public debt and decrease of spending capacity for local governments and administrations, alteration of local real estate markets, touristification of the urban environment (Di Vita and Wilson 2020).

The spread of mega-events and the rise of connected conflicts have highlighted the importance of planning and management processes in order to reduce their negative impacts (Hall 1992; Cashman and Hughes 1999), as well as to make their longer-term effects positive: the experimentation with innovative solutions for architecture, energy and mobility (Page and Connel 2012); the development of large urban transformation projects and new infrastructures (Chalkley and Essex 1999; Kassens-Noor 2012); the revitalization of entire urban districts (Wilson & Huntoon 2001; Smith 2012); the diversification of the economy and image (Clark 2008), the activation of society (Hayes and Karamichas 2011), the growth of culture and tourism (Roche 2000; Getz 2008), the turn of politics and agendas (Burbank, Andranovich, and Heying 2001, 2002; Gold and Gold 2010) of the host cities and regions.

Whilst the issue of the mega-event legacy, and its planning and management processes, have become increasingly crucial, the results achieved by or the failures of host cities and regions have become more and more contradictory (Gaffney 2013; Makarychev and Yatsyk 2016) and representative of conflicts produced by self-referential processes of globalization (Short 2008; Gruneau and Horne 2016) and 'urbanalisation' (Muñoz 2015): from the 'white elephant' left behind by the 2004 Olympics in Athens, to the disruptive acceleration and self-referential gigantism of the urban and regional development projects and processes driven by the 2014 Winter Olympics in Sochi (Müller 2014), to the radicalization of social inequalities exacerbated by the 2016 Olympics in Rio de Janeiro (Broudehoux and Sánchez 2015; Vainer 2016).

The legacy of mega-events is hard to plan, but also to assess, as it is multi-faceted and multi-scalar, and it arises differently in space and time (Evans 2020). Nevertheless, in order to avoid the increase of multi-faceted and multi-scalar conflicts (cultural, economic, environmental, political, social, spatial), connections between mega-events, dynamics of local development, and inclusive processes and tools of urban and regional governance, policy and planning are strongly required (Hiller 2012; Viehoff and Poynter 2015).

Whilst multiple conflicts have gradually led to the rise of negative perception and dissatisfaction to mega-events, with the withdrawal of several candidatures (Delaplace and Shut 2019; Kassens-Noor 2020), the more recent proliferation of smaller events (Richards and Palmer 2015; Cudny 2016) could be analysed also in relation to their capability to better combine spectacle with local issues (McMahon-Beattie and Smith 2017); that is, to reduce both the development and perception of criticalities that is more and more frequent in mega-events. Specifically, smaller events, which are more accessible not only for largest and most international cities and urban regions, but also for smallest and most peripheral towns and regions, do not usually overload infrastructures, require excessive investments, or disrupt planning processes (Roche 2000). Smaller events are not risk free, as they are another driver of festivalization and overtourism for the host cities and regions, and their impacts and effects should be better explored (Lamond and Platt 2016). However, smaller events could be considered as more scaled to the host cities and regions, involving minor places, and being less expensive to fund, easier to manage, plan and access, and more corresponding to local resources, actors and needs, with potentially positive outcomes in terms of local identity, economy, tourism, and spatial quality (Jepson and Clarke 2016; Duffy and Mair 2017).

Smaller events, which can be both temporary or periodic (Guala 2002), can be considered similar to mega-events when they are able to work as trans-scalar interfaces between local and supra-local networks (of cultural, economic, infrastructural, social, and spatial resources). However, smaller events can distinguish from mega-events according to (Di Vita and Wilson 2020):

- the capability of their formats to better reconcile the economic and political goals of repositioning of the host cities and regions with the social and environmental needs for cohesion of people and quality of places;
- the locations of their venues, which are usually placed in small, widespread, integrated and already existing facilities and spaces to reuse, as well as cultural heritage sites to enhance (being able to drive long-term processes of urban and regional regeneration), rather than in large, self-referential and disconnected facilities (which are often hard to reuse in the post-event phase).

Smaller events are more difficult to define in comparison with mega-events, and further research on their planning and management is necessary, even in the face of the growing uncertainties and demands for innovation and flexibility determined by the ongoing Covid-19 pandemic. However, smaller events could provide potential development trajectories in the face of the longer-lasting crisis of mega-events. Some of the main features of smaller events (i.e. limited costs, territorial spread of minor venues, development of temporary facilities, reuse of existing facilities and spaces, enhancement of cultural heritage) have been influencing approaches and development trajectories of contemporary mega-events: for instance, the European Capital of Culture and other similar cultural mega-events, which were born small, but have become larger according to their international organization, as well as their growing relevance and 'appetibility' for policy makers of host cities and regions (Jones 2020), with their potentials and effects of culture-led regeneration (Garcia 2005); on the contrary, the new Olympic Agenda 2020 approved by the International Olympic Committee, now at the proof on the occasion of the Summer Games of Paris 2024 and the Winter Games of Milan-Cortina 2026, that aims to innovate the mega-event format to reduce its growing unsustainability and dissatisfaction (Kobierecka and Kobierecki 2019).

In this new course of mega-events, also inspired by smaller events, unprecedented interceptions between mega-event policy and cultural heritage policy could have risks (i.e. in terms of spectacularization of cultural heritage) and potentials (i.e. in terms of development of policy for cultural heritage), which need to be analysed and verified in both theory and practice (Jones and Ponzini 2018; Ponzini et al. 2019). The attachment to place and mobilization of local communities through heritage has not been explored as an issue for mega-events so far. Accordingly, the Milan case study, that is presented in the following paragraphs, focuses on unplanned and incremental innovations, which were experimented through the side cultural events of the Expo 2015, also in connection to cultural heritage and the use of different heritage places and buildings in the historic city centre and elsewhere.

3. The context of the Milan Expo 2015, the ExpoinCittà side events, and the involvement of heritage places

This article focuses on the Milan case study arguing that the 2015 Expo was the occasion for exploring new relationships between a mega-event edition and the host city by means of side cultural events and heritage-related interventions. More generally, the context of the Milan Expo 2015, with its post-Expo and 'side Expo', is an opportunity to reflect about the contribution of mega-events to the innovation of urban agendas; the role of urban and regional policy and planning to connect the extraordinariness of mega-events to the ordinariness of urban phenomena and dynamics; the meaning and format of mega-events in a socio-economic and spatial context that has radically changed since the beginning of the 2008 global crisis (Di Vita and Morandi 2018), and that is currently changing due to Covid-19 pandemic.

The exhibitions promoted by the BIE are usually located in one, large and specifically dedicated exhibition area. In the case of the Milan Expo 2015, the official venue was located in a very accessible site (by roads, railways and airports), in a large and peri-urban area near the new Fairground, in the North-Western transect of the city. Nevertheless, Milan's strong tradition in hosting larger and smaller events inspired the development and coordination of thousands of minor collateral events (existing and new) to the World's Fair 2015. These side events to the official Expo, which were gathered through an online platform under the ExpoinCittà banner, were located in and impacted dozens of historic and modern buildings and spaces throughout both the urban core and the urban region. The most iconic of these places were the old city docks of the Navigli canal system (the so-called Darsena), that was redesigned, and the area in front of the Sforza Castle (the central Piazza Castello), that was transformed into a pedestrian area, also through temporary furniture. The Milan Expo 2015 contributed to the enhancement of the cultural heritage of the host city and region through an approach that was unprecedented in the history of the World's Fairs (Di Vita and Ponzini 2020). And this approach was also connected to the strengthening process (already ongoing) of the widespread, multi-scalar and multi-actor attractiveness of both Milan's urban core and region: that is, to the diversification process (already ongoing) of the urban and regional development trends, based on multi-faceted cultural activities and tourist flows, besides traditional manufacturing and service activities (Pasqui 2018).

The Milan Expo 2015 has its own specificity in terms of times when and places where it took place. The bidding, planning and implementation processes of the mega-event, which were promoted by a city that is historically 'pluralistic', crossed the breakout of the 2007–2008 global crisis (Bolocan Goldstein 2015, 2018; Pasqui 2015). Whilst the planning and implementation phases of the Expo 2015 were partially affected by the financial and economic downturn, the 'pluralistic' Milan exploited its strong tradition in promoting smaller but widespread events: for instance, the massive events and shopping-related events promoted by the fashion industry, as well as the Fuorisalone event (that was originally born as an alternative scene to the official Design Week, and has gradually become synergetic to the main event, strengthening its appeal and international success). According to the analysis of specific reports, websites and bibliographical references, as well as to the interviews to relevant policy makers, the research found that, in particular, the Fuorisalone event inspired the organization of thousands of side events, which were promoted by multiple actors (public and private) throughout the year of the World's Fair 2015, and frequently placed in cultural heritage sites.

The bidding phase to the Expo 2015 (2006-2007), and the following planning phase of the mega-event (2008-2011), did not focus on the multi-faceted heritage of the Milan urban core and region. However, the multi-faceted heritage of both the urban core and region was perceived as relevant to the success of the mega-event itself and its legacies. Since the development of the Candidature Dossiers, the bidding, planning, and implementation processes highlighted the relevance of both environmental and cultural heritage in the multi-scalar surroundings of the Expo site in order to strengthen the connections to the city and its region, which were considered essential to the mega-event success: from metropolitan parks, to historic centres and places, and to existing and opening cultural facilities. Specifically, the two projects of the Water Way and Land Way were promoted in order to connect the Expo site to the historical Darsena, in the main central district, by passing through the most iconic places of both the historic and contemporary city. The Water Way was planned as a canal connecting existing large parks in order to regenerate the west sector of the urban region by increasing the amount of green areas and improving the quality of landscape (for instance, through the refurbishment and reuse of the historical farms). The Land Way was planned as a permanent route across the main cultural activities and tourist attractions in the urban core.

Whilst the implementation of the Water Way was problematic and not completed, also due to protests by multiple local associations, the No-Expo committee and No-Canal campaign (Casaglia 2016), the Land Way evolved into a system of initiatives which were promoted by the Milan Municipal Administration since the election of the new Mayor, in 2011. After this political change, that corresponded to the end of political conflicts which had delayed the mega-event planning and management since the Expo awarding (in 2008), in 2012 the Milan Municipal Administration launched the City Operations Master Program in order to coordinate its multi-sectoral activities with those developed by the Expo 2015 Spa management company and other public administrations involved in the mega-event process. According to the research interview to a manager of the private consultant company *WePlan* (see Section 4), the Program was not explicitly required by the BIE, but it was inspired by the City Operations Master Program that had developed on the occasion of the Turin's Winter Olympics 2006.

According to the final version of the City Operations Master Program (including the development of side cultural events, the restoration of the former Darsena and farms, or the experimentation with the E015 digital ecosystem), the official Expo 2015 was integrated by the so-called ExpoinCittà. Partially inspired by the research *Expo Diffusa e Sostenibile* (that had been developed by Politecnico di Milano and Fondazione Cariplo; Battisti et al. 2011), ExpoinCittà was promoted by the Milan Municipal Administration and Chamber of Commerce in 2014, and it was capable of coordinating and supporting existing and new initiatives (cultural, as well as commercial and sport) by multiple stakeholders in order to exploit the tourist potentialities of the city and its region.

4. The Milan Expo 2015, side cultural events, and heritage policies: evidence and discussion

This section provides evidence and discusses the relationships (in terms of motivations, outcomes, and urban effects) between the Expo 2015, side events and cultural heritage. The research supporting this article was conducted by collecting and analysing specific reports (Arexpo 2018; Comitato di Candidatura Expo Milano 2015 2006; Comitato di Candidatura Expo Milano 2015 2007; Expo 2015 Spa 2018; ExpoinCittà 2015; Dell'Acqua et al. 2016), websites (e.g. http://en.expoincitta.com/), and bibliographical references covering the process, outcomes and legacy of the Milan Expo 2015 (Botto and Di Vita 2016; Busetti and Dente 2018), as well as the urban and regional dynamics and agendas of the Milan urban core and region (OECD 2006; Palermo and Ponzini 2012; Perulli 2016; Armondi and Di Vita 2018; Assolombarda 2018). A set of semi-structured in-depth interviews with the most important and directly involved policy makers of the planning and management process of the Expo 2015 in relation to side cultural events and heritage preservation was carried out. Specific questions on planning and implementation of both the Expo mega-event and side events, and their multi-faceted impact, effects, and legacy (cultural, political, socio-economic and spatial), were asked to (in chronological order, from November 2018 to May 2019):

(1) an officer of the Milan Municipality, Event Department;
(2) a manager of the private consultant company *WePlan. Large events, projects and more*;
(3) a manager of the Milan Municipality, Culture Council;
(4) an officer of the Italian Ministry for Cultural Heritage and Activities, Milan Superintendency;
(5) a manager of the Milan Municipality, Mayor Staff;
(6) a manager of the Milan Municipal Agency for tourism promotion, *YES!Milano*.

Furthermore, the research interpreted official data about the costs of the projects directly and not directly connected to the Expo 2015, as well as the tourist flows to the mega-event, the side events and the city, which were provided by the Milan Municipality.

According to the research interviews 1 and 4, a set of fast-track procedures, the creation of a 'one-stop-shop' for organizations to get the approval for public events, as well as an interactive digital platform (i.e. http://en.expoincitta.com/) facilitated the experimentation with the match between the supply and demand of side events with appropriate locations available, often underused, throughout the city and the region. ExpoinCittà was innovative because it provided different kinds of stakeholders with an interactive digital tool to promote their initiatives or to find available spaces for event locations. At the same time, it granted a set of procedures to validate these bottom-up proposals, as well as a set of incentives to provide the participants with free communication, and discount on taxes for temporary use of public land or for royalties on public performances.

According to the analysis conducted on this case (Di Vita and Ponzini 2020), the interactive web platform of ExpoinCittà collected 1,015 event locations in the Milan urban core and region, involving already-working event venues, as well as temporarily-used or newly-opened event sites. These event locations included:

- five new international museums (Casa del Manzoni, Fondazione Prada, MUDEC, Museo della Pietà Rondanini, and Silos Armani);
- specific cultural and exhibition spaces (both at international level, such as the Triennale's Palazzo dell'Arte, and at local level);
- other kinds of facilities and spaces (both public and private) that temporarily hosted smaller events.

The typologies of the side event locations were multi-faceted and, through a spatial model similar to that of national and European capitals of culture, they were able to enhance multiple spatial resources which are representative of Milan's cultural heritage: indeed, these event locations included amphitheatres, farms, former industrial buildings, libraries, open or public spaces, museums, theatres, villas, and other kinds of high value cloisters, porticos or buildings (source: http://en.expoincitta.com/).

45% of the 46,310 collateral events, which were officially registered in 2015 (in part existing and in part new), was placed inside the Milan historic centre. Only 10% was placed outside the Milan municipal area (ExpoinCittà 2015), thus confirming the centripetal socio-economic spatial dynamics that traditionally characterize the relationships between the Milan urban core and region, as well as the historic centre and peripheral areas.

The Expo 2015 made an important contribution to Milan's spatial re-development: not only through the implementation of several dedicated projects, but also through the acceleration of many other public and private works, already planned or under construction, with the consequent improvement of the city and region's images overall (Pasqui 2015). However, whilst the Expo site is still waiting for its post-event reuse (now slowly under development, on the background of a masterplan approved in 2018 only), the main legacy of the World's Fair 2015 can be so far recognized in complementary projects, and side cultural and heritage-based events: on the one hand, relying on the development of new transport and digital infrastructures, already included in the Bidding Dossiers; on the other hand, leading to the acceleration of urban regeneration processes, in particular in the central and creative districts strongly involved by the ExpoinCittà (i.e. the Navigli, Porta Genova and via Tortona neighbourhoods, where also the renovated Darsena is located, or the Cadorna, Brera, and Porta Garibaldi areas, around the Sforza Castle) that was not planned at the beginning of the Expo process, but experimentally and incrementally developed by enhancing existing projects and places, policies, and socio-economic dynamics.

In comparison with the official Expo site and infrastructures, the costs of the City Operations Master Program (including the ExpoinCittà, and the renewal of the Darsena and the area in front of the Sforza Castle) were significantly lower: only 39 million euro (generating 11 million visitors), in contrast with the total expenditure of 14,780 million euro for the entire Expo processes and projects (generating 21.5 million visitors).

According to the research interview to a manager of the private consultant company *WePlan*, despite the Expo theme (*Feeding the planet, energy for life*), the promotion of the international Urban Food Policy Pact and the development of the local Milan Food Policy on the occasion of the Expo 2015, Milan and the World's Fair 2015 have not been able, yet, to really renovate the city image and dynamics in relation to its local agricultural surroundings. Consequently, the planned re-development of Milan in relation to the local environmental and agricultural resources of its surroundings (Expo 2015 Spa 2009) has been so far episodic whereas, before the Covid-19 pandemic, the urban core, its historic centre and cultural heritage have become more and more attractive for multi-faceted economic activities and tourist flows.

All interviewees confirmed that thousands of widespread locations of side events produced more relevant urban outcomes than the Expo official projects themselves, especially by looking at the historic city fabric and to places like the pedestrianized square in front of the Sforza Castle, as well as the renovated old Navigli docks. These side-events and their effects did not contribute to develop a new, explicit and long-term spatial vision for the urban core and region, but rather to change and adapt the public use and general meaning of several cultural heritage sites and spaces, and to develop a new image of Milan as an international destination of cultural tourism. However, growing tourist flows (from 3.3 million arrivals in 2007, to 5.3 million in 2015 and 5.5 million in 2017 – source: Milan Municipality) and renovated tourist services contributed to the festivalization of the historic city centre and overtourism in multiple heritage sites, exacerbating the existing socio-spatial disparities and conflicts between central and peripheral areas (Cucca and Ranci 2017), increasing the traditional divide between the urban core and region (Garofoli 2016), and fuelling further protests against the event (Casaglia 2016).

In sum, ExpoinCittà has shown that contemporary mega-events can still be capable of activating multiple resources (funding, institutional and professional expertise, projects and places), as well as sparking different interpretations and uses of multiple spaces in cities and regions, well beyond the official location and narrative of the mega-event itself: and in this scenario, cultural side events and heritage can play a relevant role. The experimental and incremental initiative of the Expo side events, that mixed top down and bottom up approaches, was able to systematically detect, map, brand and exploit available facilities and places, thus providing new development models and trajectories for cities and regions, as well as mega-events. Through its innovative open-data collection, it allowed Milan to foster and spread thousands of cultural events to a scale that was not neither envisioned nor requested by the Expo format promoted by the BIE. At the same time, it significantly and positively contributed to the development of the effects and legacy of the official Expo 2015. ExpoinCittà increased the availability of spaces and fuelled economic interests for more and more events, and it achieved these outcomes also by assigning a relevant role to cultural heritage. Nevertheless, risks of festivalization and touristification of the city and its region have threatened these results: on the one hand, according to the high concentration of cultural and entertainment activities in the historic centre, the erosion of public space, the overpressure and limits in the appreciation of heritage sites, and the growing disparities with peripheries and other excluded areas; on the other hand, according to the socio-economic and spatial shocks determined by the Covid-19 pandemic.

5. Urban agenda between political consensus and impact mitigation

One can derive important urban agenda considerations from the evidence of this article. The specific competences and powers of the Expo and post-Expo planning and management companies (respectively, Expo 2015 Spa and Arexpo Spa), as well as the procedures bypassing the ordinary management of public works, led the Expo site and the related transport infrastructures to be planned and coordinated mainly through a rigid top-down approach. On the contrary, the complementary city cultural programmes and projects were experimentally and incrementally developed through the involvement of multiple public and private actors, being the ExpoinCittà web platform provided by the Milan Municipality and Chamber of Commerce the means for matching supply and demand of side events.

The reasons of this innovation can be led back to the multiple problems and delays in the official mega-event planning and management, which were also connected to both the 2008–2011 political conflicts related to the definition of the mega-event plans and governance (Botto and Di Vita 2016), and the decrease of available public and private funding determined by the impact of the global crisis, with risks of mega-event downsizing: for instance, the cost of Expo 2015 Spa decreased from 4,120 million euro estimated by the 2007 Bidding Dossier (Comitato di Candidatura Expo Milano 2015 2007), to 2,196 million euro registered at the end of the mega-event celebration (Expo 2015 Spa 2018).

Even though it is difficult to separate the event legacies and their specific reasons from the ongoing processes and trends of urban change, the spread of the Expo side events made an important contribution to the official exhibition success (Expo 2015 Spa 2018), and to the long-lasting processes of urban re-development and repositioning in relation to both local surroundings and world networks that it intercepts (Bolocan Goldstein 2015 and 2018). Furthermore, the BIE itself has appreciated the innovation and quality of the ExpoinCittà experimentation, that is the importance of integrating its traditional exhibition platforms, usually working as urban enclaves, with a system of collateral minor events spreading across the host city and region, and exploiting cultural heritage and other local resources. For this reason, the BIE has indicated the Milan ExpoinCittà as a best practice to include within the future Expo agreements to sign with the next host cities and regions (ExpoinCittà 2015).

The Expo side events made an important contribution to the development of multi-sectoral and multi-scalar governance, that was partially based on existing expertise and partially stimulated by the mega-event. For instance, according to the research interview to an officer of the Italian Ministry for Cultural Heritage and the Arts, the Milan Superintendency was involved in the development of ExpoinCittà in order to give its approval to and monitor the numerous projects and events placed inside the local cultural heritage. Consequently, the Milan Superintendency intensified the activities of a joint inter-institutional committee with the Milan Municipal Administration without modifying its ordinary procedures, but exploiting an expertise already consolidated in a city hosting several ephemeral events per year. Within the complexity of the planning and governance of the mega-event and the connected side events, the existing expertise in approving and monitoring hundreds of temporary and ephemeral events, which were often placed in historic spaces and buildings, contributed to strengthening and casting the image of a very dynamic city at multiple levels, both local and supra-local.

According to the research interviews to a high-level officer of the Milan Municipality, Event Department, and a manager of the Municipal Agency for tourism promotion, *YES! Milano*, ExpoinCittà has also led to the establishment of the Milan Municipal Administration's *Sportello Unico Eventi (SUEV)*, that has simplified the procedures of event authorization and monitoring, and the Municipal Agency *YES!Milano* itself, that has reorganized the thousands of city events into thematic weeks, all year long: from top-down events (i.e. Piano City) to bottom up events (i.e. Design Week); from specific networks' events (i.e. Fashion Week) to coordinated events (i.e. Darsena-Navigli events). This is part of the side Expo legacy, although the innovative services and approach provided by the ExpoinCittà web platform have been recently reduced by the new Municipal Agency, that is limiting its activity to the promotion of the city brand.

During the interviews, two managers of the Milan Municipality, at the Culture Council and Mayor Staff, respectively, confirmed that the goals of ExpoinCittà were multiple, while enhancing Milan's long-term tradition and expertise in hosting smaller but widespread events: for instance, supporting a mega-event in crisis in terms of budget cuts and work delays during the planning and implementation phase (2012-2014); enlarging the city and region's connections and participation to the mega-event; increasing the consensus to the Expo 2015. At the same time, one of the main outcomes of ExpoinCittà has been the reinforcement of both the Milan public institutions and private actors' expertise and political orientation towards (smaller and larger) events.

It is important to highlight the fact that the former CEO of the Expo 2015 Spa planning and management company was elected City Mayor, in 2016. A pro-growth, event-fuelled, and tourism-based approach to city planning and management has been further supported from the above-mentioned foundation of the Municipal Agency *YES!Milano*. The successful bid to the Winter Olympics of Milano-Cortina 2026 confirmed the trend and the importance of mega-events in the city agenda despite evident side effects. Furthermore, in the post-Expo the local authorities have been also attempting at facing growing processes of urban festivalization and touristification: on the one hand, they aim to contain the heritage pressures by reducing the use for temporary events of the most famous places in the historic centre; on the other, they aim to decrease the congestion of central areas and broaden the ongoing urban renaissance to peripheral districts, which have unexpressed potentialities to exploit or to contrast marginalization trends, by incentivizing the spread of smaller events in other less known sites across both the city centre and peripheral areas. This new approach, experimented during the post-Expo phase, has been directly driven by the new political agenda, that has been set by the current City Council. The 'Piano Perifierie' of the Milan Municipality and the programme 'Welfare Metropolitano' of the Milan Metropolitan City have introduced goals, strategies, and actions to reduce gaps in the urban core and region, by containing pressure on the historic centre and redistributing wealth to the outskirts: for instance, to bridge the festivalization and touristification of urban spaces with social inclusion, also by fostering cultural, leisure and sport events in available facilities and public spaces located in peripheral areas.

As it is too early to assess the Winter Olympics of Milano-Cortina 2026 (see the official websitehttps://www.milanocortina2026.org/) and to fully understand the implications and impact of the Covid-19 pandemic over events and heritage places, one should wait and see if the approach and solutions experimented by ExpoinCittà will be adopted and will be capable of building consensus for this next mega-event.

6. Conclusions

The Milan case study shows that mega-events can benefit from smaller events and heritage sites, and related processes and projects, which could unlock new opportunities and support political orientation towards events and related policies. At the same time, the case study shows that deadlines and additional resources, which are typically given by mega-events, can become a clear opportunity for enhancing smaller events and heritage sites, and related processes and projects of preservation and re-use (Jones and Ponzini 2018; Jones 2020), even when they are not explicitly requested by an inter-governmental organization like the BIE. However, this approach may be a double-edged sword from the political point of view. On the one hand, ExpoinCittà mobilized thousands of actors and welcomed millions of visitors creating wealth and consensus for event-fuelled strategies. On the other hand, the Expo 2015 and its side events contributed to accelerating negative phenomena of festivalization and touristification of the Milan historic centre, with connected problems of overtourism, erosion of public spaces, and overpressure of cultural heritage sites. Furthermore, they exacerbated the already ongoing dualism of urban and regional development trends between the city centre and peripheral areas.

Both the pros and cons of the Milan case study highlight the urgent demand for re-interpretation and re-consideration of outdated relationships between (larger and smaller) events, hosting cities and regions, and traditional development models of urban and regional planning and governance (Müller 2015; Gruneau and Horne 2016; Lauermann 2019), where tensions between cultural heritage and socio-spatial disparities can be focal (Garcia and Cox 2013; Richards and Palmer 2015; Ponzini and Jones 2020). These relationships become more and more crucial in the face of the growing uncertainties that the Covid-19 pandemic projects on events, heritage policies and our urban life more generally. These points urgently require new research and policy debate for a city like Milan that is planning for the 2026 Winter Olympics, as well as for many other cities and regions interpreting events as a strategy for growth.

Acknowledgments

This article is dedicated to the memory of Corinna Morandi (Politecnico di Milano, DAStU), with whom I shared my research interests and activities on the relationships between mega-events, cities and planning since the beginning of my career. I would like to thank Davide Ponzini (Politecnico di Milano, DAStU) for fruitful cooperation as part of the 'HOMEE' research project. I would also like to thank all the officers and managers of the Milan Municipality, the Milan Superintendency and the WePlan private consultant company for providing precious data, information and reflections for the implementation of the research project. I am grateful to two anonymous referees for their comments to the earlier versions of this article.

Funding

Sections of this paper are based on and supported by the research conducted within the JPI Cultural Heritage research project 'HOMEE – Heritage Opportunities/threats within Mega-Events in Europe', with a grant issued by the Italian Ministry of University and Research.

References

Arexpo Spa. 2018. *MIND – Milan Innovation District. Il parco della scienza, del sapere e dell'innovazione*. Presentazione.

Armondi, S., and S. Di Vita, eds. 2018. *Milan. Productions, Spatial Patterns and Urban Change*. London: Routledge.

Assolombarda. 2018. *Milano Scoreboard* 2018. Report. (Accessed Mar 10, 2019. https://osservatoriomilanoscoreboard.it/sites/default/files/2018-06/Osservatorio-Milano-2018-ENG.pdf.)

Battisti, E., F. Battisti, S. Di Vita, and C. Guerritore. 2011. *Expo Diffusa e Sostenibile*. Milano: Unicopli.

Bolocan Goldstein, M. 2015. "Post-Geographical Scenarios." *Urbanistica* 155: 118–122.

Bolocan Goldstein, M. 2018. "Urban Regionalization and Metropolitan Resurgence. Discontinuity and Persistence of a Spatial Dialectic." In *Milan. Productions, Spatial Patterns and Urban Change*, edited by S. Armondi, and S. Di Vita, 36–47. London: Routledge.

Botto, I., and S. Di Vita. 2016. *Oltre L'Expo 2015. Tra Dimensione Ordinaria e Straordinaria Delle Politiche Urbane*. Roma: Carocci.

Broudehoux, A., and F. Sánchez. 2015. "The Politics of Mega-Event Planning in Rio de Janeiro: Contesting the Olympic City of Exception." In *Mega-Event Cities: Urban Legacies of Global Sport Events*, edited by V. Viehoff, and G. Poynter, 109–122. Farnham: Ashgate.

Burbank, M., G. Andranovich, and C. Heying. 2001. *Olympic Dreams: The Impact of Mega-Events on Local Politics*. London: Lynne Rienner Publisher.

Burbank, M., G. Andranovich, and C. Heying. 2002. "Mega-Events, Urban Development, and Public Policy." *Review of Policy Research* 19 (3): 179–202.

Busetti, S., and B. Dente. 2018. *Expost. Le conseguenze di un grande evento*. Bologna: Il Mulino.

Casaglia, A. 2016. "Territories of Struggle: Social Centres in Northern Italy Opposing Mega-Events." *Antipode* 50 (2): 478–497.

Cashman, R., and A. Hughes, eds. 1999. *Staging the Olympics. The Event and its Impacts*. Sydney: University of New South Wales Press.

Chalkley, B., and S. Essex. 1999. "Urban Development Through Hosting International Events: a History of the Olympic Games." *Planning Perspective* 14 (4): 369–394.

Clark, G. 2008. *Local Development Benefits from Staging Major Events*. Paris: OECD.

Comitato di Candidatura Expo Milano. 2015 (2006). *Proposta di Candidatura Della Città di Milano a Ospitare L'Esposizione Universale del 2015*. Milano: Comitato di Candidatura Expo Milano 2015.

Comitato di Candidatura Expo Milano. 2015 (2007). *Dossier di Candidatura Della Città di Milano a Ospitare L'Esposizione Universale del 2015*. Milano: Comitato di Candidatura Expo Milano 2015.

Cucca, R., and C. Ranci. 2017. *Unequal Cities: The Challenge of Post-Industrial Transition in Times of Austerity*. London: Routledge.

Cudny, W. 2016. *Festivalisation of Urban Spaces: Factors, Processes and Effects*. Cham: Springer.

Delaplace, M., and P. O. Shut, eds. 2019. *Hosting the Olympic Games: Uncertainty, Debates and Controversy*. London: Routledge.

Dell'Acqua, A., G. Morri, E. Quaini, and A. Airoldi. 2016. *L'indotto di Expo 2015. Un'analisi di Impatto Economico al Termine Dell'evento*. Milano: SDA Bocconi. Camera di Commercio di Milano and Expo 2015 Spa.

Di Vita, S., and C. Morandi. 2018. *Mega-Events and Legacies in Post-Metropolitan Spaces. Expos and Urban Agendas*. Basingstoke: Palgrave Macmillan.

Di Vita, S., and D. Ponzini. 2020. "Milan Expo 2015. The Spread of Cultural Events in Historic Places and Beyond." In *Mega-Events and Heritage: The Experience of Five European Cities*, edited by D. Ponzini, Z. Jones, F. Bianchini, J. Georgi, and J. Sanetra-Szeliga, 60–101. Krakow: International Cultural Centre.

Di Vita, S., and M. Wilson, eds. 2020. *Downsizing Spectacle: Planning and Managing Smaller (Urban) Events*. London: Routledge (In press).

Duffy, M., and J. Mair. 2017. *Festival Encounters: Theoretical Perspectives on Festival Events.* London: Routledge.
Evans, G., eds. 2020. *Mega-Events: Placemaking, Regeneration and City-Regional Development.* London: Routledge.
Expo 2015 Spa. 2009. *Conceptual Masterplan.* Milano: Expo 2015 Spa.
Expo 2015 Spa. 2018. *Expo Milano 2015. Report ufficiale.* La sfida dell'Italia per un'esposizione universale innovativa. Milano: Expo 2015 Spa.
ExpoinCittà. 2015. *2015 Milano a place to be. L'esperienza di ExpoinCittà.* Milano: ExpoinCittà.
Gaffney, C. 2013. "Between Discourse and Reality: The Un-Sustainability of Mega-Event Planning." *Sustainability* 5: 3926–3940.
Garcia, B. 2005. "De-constructing the City of Culture: The Long Term Cultural Legacies of Glasgow 1990." *Urban Studies* 42 (5/6): 1–28.
Garcia, B., and T. Cox. 2013. *European Capitals of Culture: Success Strategies and Long-Term Effects.* Luxembourg: European Parliament Committee on Culture and Education.
Garofoli, G. 2016. "Regione Urbana Milanese: una Metropolizzazione Povera." *Argomenti e Contributi* 15: 36.
Getz, D. 2008. "Event Tourism: Definition, Evolution, and Research." *Tourism Management* 29 (3): 403–428. Doi: 10.1016/j.tourman.2007.07.017.
Gold, J., and M. Gold, eds. 2010. *Olympic Cities. City Agendas, Planning and the World's Games, 1896-2012.* London: Routledge.
Gruneau, R., and J. Horne, eds. 2016. *Mega-Events and Globalization. Capital and Spectacle in a Changing World Order.* London: Routledge.
Guala, C. 2002. "Per una tipologia dei mega eventi." *Bollettino della Società Geografica Italiana* 7 (4): 1–25.
Hall, C. 1992. *Hallmark Tourist Events. Impacts, Management and Planning.* London: Belhaven.
Hayes, G., and J. Karamichas, eds. 2011. *Olympic Games, Mega-Events and Civil Societies.* Basingstoke: Palgrave MacMillan.
Hiller, H. 2012. *Host Cities and the Olympics: an Interactionist Approach.* London: Routledge.
Jepson, A., and A. Clarke, eds. 2016. *Managing and Developing Communities, Festivals and Events.* Basingstoke: Palgrave Macmillan.
Jones, Z. 2020. *Cultural Mega-Events: Opportunities and Risks for Heritage Cities.* London: Routledge.
Jones, Z., and D. Ponzini. 2018. "Mega-events and the Preservation of Urban Heritage: Literature Gaps, Potential Overlaps, and a Call for Further Research." *Journal of Planning Literature* 33 (4): 433–450.
Kassens-Noor, E. 2012. *Planning Olympic Legacies. Transport Dreams and Urban Realities.* London: Routledge.
Kassens-Noor, E. 2020. *Los Angeles and the Summer Olympic Games: Planning Legacies.* Cham: Springer.
Kobierecka, A., and M. Kobierecki. 2019. "The International Olympic Committee's Struggle Against Growing Gigantism of the Olympic Games." In *Hosting the Olympic Games: Uncertainty, Debates and Controversy,* edited by M. Delaplace, and P. O. Shut, 48–60. London: Routledge.
Lamond, I., and L. Platt, eds. 2016. *Critical Event Studies: Approaches to Research.* Basingstoke: Palgrave Macmillan.
Lauermann, J. 2019. "The Urban Politics of Mega-Events." *Environment and Society* 10 (1): 48–62.
Makarychev, A., and A. Yatsyk, eds. 2016. *Mega Events in Post-Soviet Eurasia: Shifting Borderlines of Inclusion and Exclusion.* Basingstoke: Palgrave MacMillan.
McMahon-Beattie, U., and K. Smith. 2017. *The Future of Events & Festivals.* London: Routledge.
Muñoz, F. 2015. "Urbanalisation and City Mega-Events. from 'Copy & Paste' Urbanism to Urban Creativity." In *Mega-event Cities: Urban Legacies of Global Sport Events,* edited by V. Viehoff, and G. Poynter, 11–22. Farnham: Ashgate.
Müller, M. 2014. "After Sochi 2014: Costs and Impacts of Russia's Olympic Games." *Journal of Eurasian Geography and Economics* 55 (6): 628–655.

Müller, M. 2015. “The Mega-Event Syndrome: Why So Much Goes Wrong in Mega-Event Planning and What To Do About It.” *Journal of the American Planning Association* 81 (1): 6–17.
OECD. 2006. *OECD Territorial Reviews: Milan, Italy*. Paris: OECD.
Page, S., and J. Connel, eds. 2012. *The Routledge Handbook of Events*. London: Routledge.
Palermo, P., and D. Ponzini. 2012. “At the Crossroads Between Urban Planning and Urban Design: Critical Lessons from Three Italian Case Studies.” *Planning Theory & Practice* 13 (3): 445–460.
Pasqui, G. 2015. “Expo 2015 and Milan: Intertwined Stories.” *Urbanistica* 155: 106–109.
Pasqui, G. 2018. “The Last Cycle of Milan Urban Policies and the Prospects for a New Urban Agenda.” In *Milan. Productions, Spatial Patterns and Urban Change*, edited by S. Armondi, and S. Di Vita, 133–144. London: Routledge.
Perulli, P. 2016. *The Urban Contract: Community, Governance and Capitalism*. London: Routledge.
Ponzini, D., and Z. Jones. 2020. “Emerging Issues, Opportunities and Threats and How to Deal with Them.” In *Mega-Events and Heritage: The Experience of Five European Cities*, edited by D. Ponzini, Z. Jones, F. Bianchini, J. Georgi, and J. Sanetra-Szeliga, 316–343. Krakow: International Cultural Centre.
Ponzini, D., Z. Jones, F. Bianchini, E. Tommarchi, J. Georgi, E. Dova, A. Sivitanodou, et al. 2019. *Literature Review of Mega-Events Addressing Cultural Heritage Issues*. Krakow: International Cultural Centre.
Richards, G., and R. Palmer. 2015. *Eventful Cities: Cultural Management and Urban Revitalisation*. London: Routledge.
Roche, M. 2000. *Mega-Events and Modernity*. London: Routledge.
Short, J. 2008. “Globalization, Cities and the Summer Olympics.” *City: Analysis of Urban Trends, Culture, Theory, Policy, Action* 12 (3): 321–340.
Smith, A. 2012. *Events and Urban Regeneration. The Strategic use of Events to Revitalise Cities*. London: Routledge.
Vainer, C. 2016. “Mega-Events and the City of Exception: Theoretical Explorations of the Brazilian Experience.” In *Mega-Events and Globalization. Capital and Spectacle in a Changing World Order*, edited by R. Gruneau, and J. Horne, 97–112. London: Routledge.
Viehoff, V., and G. Poynter, eds. 2015. *Mega-Event Cities: Urban Legacies of Global Sport Events*. Farnham: Ashgate.
Wilson, M., and L. Huntoon. 2001. “World’s Fairs and Urban Development. Lisbon and Expo 98.” *International Review of Comparative Public Policy* 12: 373–394.

Culture and heritage as a means to foster quality of life? The case of Wrocław European Capital of Culture 2016

Joanna Sanetra-Szeliga

ABSTRACT

Wrocław European Capital of Culture 2016, with its 'Spaces for Beauty' slogan, serves to show that cumulating financial resources and human effort to produce a large number of projects and initiatives in the field of heritage provides an opportunity to improve the quality of life for residents. This paper discusses two aspects of quality of life: access to and participation in culture as well as relations with urban space and built heritage. Both impact numerous areas of our lives, such as individual satisfaction, self-development, identity building, sense of belonging, sense of security, aesthetic pleasure, entertainment, learning, socialization of individuals and building social capital. This paper aims to bridge the gap between research on mega-events and the discussion on heritage, offering insights into the potential use of heritage as a development mechanism.

1. Introduction

This paper is intended to draw attention to one of the rarely studied aspects of the European Capital of Culture (ECoC) initiative: heritage and quality of life. Its goal is to analyse heritage-related projects organized in Wrocław in 2016, in an attempt to discuss their potential contributions to quality of life. Moreover, it addresses the research gap, identified by Ponzini (2019), between a growing body of studies on cultural mega-events, especially ECoCs (including Richards and Palmer 2010; García and Cox 2013; Giovanangeli 2015) and a cultural heritage debate, focusing here especially on one aspect of the holistic approach to its socio-economic impact – quality of life (Giraud-Labalte 2015). Although a number of authors have analysed the impact of mega-events (see, for example, Langen and García 2009; Stone and Sharpley 2011; Herrero et al. 2006), they tend to focus on either social or economic aspects of the potential impact. To my knowledge, no one has tried to analyse the relationship between a mega-event and heritage-led quality of life. In line with other arguments for the ability of cultural heritage to contribute to different aspects of quality of life, this article aims to show that the ECoC 2016, as a cultural mega-event (Jones 2020) bringing a concentration of financial resources, human effort, projects and initiatives, was a framework potentially catalysing socio-economic change. I elaborate here on the existing evaluations of short-term effects and opportunities the discussed heritage projects bring, keeping in mind that the long-term effects of ECoC 2016 are yet to be observed and analysed. It should be emphasized that, due to the lack of 'ex ante' analyses on heritage-related aspects of quality of life in Wrocław and any evaluation of this area during ECoC, this 'ex post' analysis is limited and rather more based on deductive reasoning.[1]

The analysis is based on a comprehensive analysis of Wrocław 2016 bid documents, projects, reports and other relevant municipal documents. It also draws on a media content analysis accompanied by a series of 15 in-depth interviews with Wrocław 2016 stakeholders (representatives of the municipality, the ECoC implementing agency, project organizers, journalists, local activists and experts) and two focus group interviews (one with people directly involved in managing the ECoC 2016 implementation – the ECoC agency and the municipality – and the other with outside experts). Starting with Wrocław's mega-events-led development strategy, the article moves on to discuss the ECoC heritage-related projects as contributors to quality of life. The article focuses on heritage-related projects, selected from the ECoC program on the basis of experts' assessment[2], since the ECoC 2016 narrative did not use the term 'heritage'.[3] Based on my previous research (Sanetra-Szeliga 2016), I argue that the socio-economic impact of cultural heritage goes beyond commonly acknowledged contributions to tourism and city branding. The article concentrates rather on two aspects of quality of life: access to and participation in culture and relations with urban space and built heritage. Here, the analysis draws on the research of numerous authors dealing with heritage impact on quality of life (e.g. Davies, Whimster, and Clayton 2009; Fujiwara, Kudrna, and Dolan 2014; Matarasso 1997; Blessi 2014), analysed in detail by Sanetra-Szeliga and Jagodzińska (2017).

2. Mega-events as a development strategy in Wrocław

Since the ECoC 1990 in Glasgow, the cultural mega-event has increasingly been used by host cities as a mechanism for urban regeneration and transformation (e.g. García 2004; Paris and Baert 2011). Central and Eastern Europe post-communist cities used it as an attempt to prove their 'Europeanness' by exploring European values and 'catching up' with European standards (Lähdesmäki 2014, 483). Hosting mega-events was a way 'to demonstrate to the rest of the world that the cities and countries once behind the Iron Curtain have at long last arrived in global modernity' (Müller and Pickles 2015, 121). This modernity in many cases meant finally upgrading their transport, communication, or energy infrastructure, while regeneration or renewal schemes were given second priority (contrary to what happens in many Western cities) (124). Major investments in buildings tipped to become iconic and a belief in the 'Bilbao effect' (Gómez 1998) are still much more common in Europe's east than in its west, where decision-makers seem to have become more aware of the 'white elephant' problem (Smith 2012, 44–45) regarding giant infrastructural investments.

In Wrocław, one of the major ideas for development in the 1990s and 2000s was to host a mega-event that would serve as a 'leap forward' in terms of social and economic development, bringing in capital, tourists and opportunities for social change. The first attempt to attract a mega-event (the UEFA European Championships 2012) succeeded in 2005. It was then followed by two failed bids for the EXPO, in 2010 and 2012. Hosting Euro 2012 aligned with the above-mentioned pattern of major investments in the city's infrastructure (a new football stadium, the 35-km highway ring-road, the regeneration of the public transport hub, a new airport terminal) (Filip 2012) and 'strengthening the feeling that we are part of the civilized world' (cg 2012). It improved Wrocław's international recognition and furthered its promotion, bringing also substantial income to the city – 325,000 tourists and football fans spent ca. EUR 59 M (Deloitte 2012) in the city during the games.

'It is always good to have something that can focus effort and highlight further goals', one of the high-level Wrocław officials claims (WroMun01). Consequently, mega-events have been used as such tools or mechanisms several times since Euro 2012. The ECoC candidacy that followed was motivated by this belief, as was a subsequent mega-event organized in the city: the 2017 World Games. The municipality's confidence that internationally recognized events and titles matter for the city's image and development was also mirrored in Wrocław's term as UNESCO World Book Capital 2016 (in parallel to the ECoC) and its accession to the UNESCO Creative Cities Network in 2019 (after having been awarded the title of UNESCO Creative City of Literature).

3. ECoC 2016 catalysing the change in quality of life

The Strategy 'Wroclaw in the Perspective 2020 plus' (Galar 2006), in force during ECoC 2016 bidding and implementation, recognizes a number of challenges that experts (e.g. Hermant-de Callataÿ and Svanfeldt 2011; Hall 2014) list as urgent in twenty-first-century cities. These include social issues (e.g. alienation of individuals, the need to foster the city's symbolic community and self-fulfilment of its residents) and spatial challenges (e.g. striving for well-formed and friendly public spaces). As major transport and communication infrastructure investments had already been implemented for Euro 2012, the ECoC bid and its implementation could focus more on socio-cultural change. Wrocław used the ECoC as an opportunity to reconsider the role of culture as a socio-economic resource and think more in terms of a co-creative process. The conviction was that the city's earlier growth had been reliant on quantitative resources that were largely exhausted. It was time to reorient towards 'qualitative change', including 'mobilizing civic, social, professional, and creative activity' (Chmielewski 2011, 7) – and culture, explicitly, was perceived as the city's chance for development (6).

This social development, according to the ECOC slogan 'Spaces for Beauty', was to be accomplished by establishing 'the presence of beauty in social and personal life' and creating 'spaces within which to restore the presence of beauty in public life and in daily habits' (13). During the implementation phase, the following detailed objectives were extrapolated from these aims: increasing access to culture and encouraging participation, renovation of a number of buildings, districts and urban areas (the Reclaiming Beauty program), and creating new cultural infrastructure (the New Spaces for Beauty program). Based on the extensive literature review (Sanetra-Szeliga 2017), I argue that these can be linked with quality of life, as discussed below. The 425 ECoC projects implemented in 2016 resulted in over 2000 major cultural events (Banaszak 2017a, 12) organized around these objectives.

Quality of life is an important element of successful urban development and attracting people (residents and tourists as well as investors). 'The ability of a city to create an attractive environment so as to enhance the quality of living for the residents' (Ni

2012, 14), which is one of the important determinants in decisions on where to locate businesses and live (Rogerson 1999), recurs in many definitions of urban competitiveness as well as in the contemporary approach to socio-economic development. Here, quality of life is understood as 'the extent to which objective human needs are fulfilled in relation to personal or group perceptions of subjective well-being' (Constanza 2008, 18). An analysis of research and rankings related to quality of life (Sanetra-Szeliga and Jagodzińska 2017) reveals a lot of quality of life stimuli in which cultural heritage can have a role to play, including psychological well-being, life satisfaction, subjective health assessment, sense of security, sense of belonging, social integration and identity building. Table 1 presents the potential of culture and cultural heritage with regards to raising quality of life and factors that affect urban quality of life.

The following sections will present two complementary parts that affect quality of life. The first one could be called 'the software' and refers to access to cultural offer and participation in it, while the other, 'the hardware', deals with the physical urban fabric and built heritage.

3.1. Access and participation in culture

An estimated 5.2 M[4] people (Banaszak 2017a, 170) took part in ECoC projects throughout the year. There were a wide variety of formats to choose from, including mass events (e.g. the Flow Quarter), special editions of well-known Wrocław festivals (e.g. Summer in the White Stork Synagogue festival), a number of one-off special projects from several cultural domains of varying sizes, and projects focused on actively engaging local residents.

Pluta (2017, 7) refer to the ECoC 'a kind of mass social intervention – a type of experiment in which the residents of the city and the region actively participate'. This active participation was particularly evident in the case of the microGRANTS scheme. Local NGOs, artists and other individuals could apply for support, which consisted in a small cash injection, as well as organizational, administrative and logistical assistance. The idea was to involve residents in the organization of the ECoC, build 'goal communities', inspire people to create and organize events, build their identity and offer education in leadership competences. In 2014–2016, 405 microprojects were supported (Dolińska, Kajta, and Niedźwiecka-Iwańczak 2017a, 12, 21, 23). 10% of microgrants financed cultural heritage-led initiatives, some of which were rooted in the history and uniqueness of a given neighbourhood, with a goal of reinforcing the sense of place and exploiting the potential to bring people together around a common historical background. A good example here would be the project Gateway to the Future. This involved the rediscovery of a forgotten part of the city (Ratyń) by collecting memories, stories, photos and documents related to the neighbourhood and using them to integrate residents through various activities, such as staging local legends or organizing a photography competition.

The authors of the scheme's evaluation argue that it has succeeded in overcoming a general attitude of 'Why bother?' and 'I don't feel like it' and changed it more into 'Maybe we want to do it? Let's try' (Dolińska, Kajta, and Niedźwiecka-Iwańczak 2017a, 38). The scheme also proved that the residents were interested in their city, they wanted to know and understand it better, and sensitivity towards saving its history and people from oblivion was awoken. The formats used to achieve that might not have been very sophisticated, such as 'a simple walk through the neighborhood,

Table 1. Culture and cultural heritage as components of urban quality of life. Selected issues.

	Quality of life subparts	Potential of culture	Urban quality of life factors
Health	Subjective health assessment	Active and passive participation in culture influences one's health self-assessment	Access to culture
	Feeling of stress	Active and passive participation in culture can reduce stress thanks to detachment from everyday life, feeling of freedom, and the possibility of discharging negative energy. A properly designed public space affects the mental state of a person.	Well-designed and accessible public spaces
Life satisfaction	Education and learning	Possibility to develop hobbies and interests by participating in various types of cultural activities	Access to a diverse cultural offer as a form of education and development
	Self-assessment of satisfaction deriving from social interactions	Culture as a relational good can be a way to deepen interpersonal relations	Cultural institutions as third places[a]
	Self-assessment of spending one's free time	Active and passive participation in culture is a creative consumption which, thanks to its diversity and the production of new goods and services, can combat the monotony of everyday life and give a sense of personal development.	Access to a diverse cultural offer
Security	Feeling of security	A properly preserved and promoted cultural heritage (tangible and intangible) gives a sense of belonging and builds identity. Historic urban fabric affects the mental stability of an individual and their sense of security. A similar function is performed by a well-designed urban space. Culture also influences the building of social capital, which can constitute a sense of security.	Taking care of historic buildings and good quality of public space, architecture and urban planning Cultural activities in degraded areas, with groups at risk of exclusion and undertaking criminal activities such as vandalism, hooliganism, or petty crime Cultural activities focused on getting to know the Other, intercultural dialogue and integration of local communities

[a]'Third places' are spaces easily accessible, relaxing, and pleasant, where one meets friends and makes new acquaintances. They are important for building identity, feeling of belonging, strengthening social capital (Oldenburg 1989).

Source: Own elaboration of the literature review, in particular: Sanetra-Szeliga (2017), Psatha, Deffner, and Psycharis (2011), Richards (2014), Blessi (2014) and Matarasso (1997).

showing its historic background, opening the district, to some extent, to the residents' (Dolińska, Kajta, and Niedźwiecka-Iwańczak 2017a, 39). However, what mattered was that the goal of active participation of residents was achieved.

3.1.1. Culture closer to residents

Providing access to culture means not only preparing an interesting cultural offer in traditional cultural institutions and their venues but also outdoors and in public spaces. A number of projects implemented this idea, including the Flow Quartet, the Backyard Door project and the ECoC parks.

This process arguably went hand in hand with the ECoC organizers' desire to decentralize the cultural life of the city somewhat. The opening ceremony of Wrocław 2016, the Awakening (January 15–17, 2016), can be seen as a way to engage different parts of the city in the mega-event. The program consisted of over a hundred events, culminating in parades starting from four different corners of Wroclaw and gradually approaching the meeting point at the market square (see Figure 1).

This was the most described and most commented ECoC event in both traditional and social media, where it received contradictory reviews, ranging from criticism of serious organizational and communication shortcomings to admiration for the enormous turnout - 120,000 people (Banaszak 2017a, 168) - and the finale of the event on the Market Square. Relating to the history of the city evoked emotions that bonded the people of Wroclaw in a joint experience during the parades. As Kajdanek et al. point out, the crowd was animated in a common ritual. However, the Awakening itself was evaluated relatively poorly due to the loss of momentum (2017, 9).

Heritage-related projects were largely concentrated in the historic center, rich in historical architecture (see Figure 2). Figures 3 and 4 well illustrate the dissemination of projects across the city space. While regular cultural institutions hosted just a few of the heritage-related initiatives, a significant proportion of the heritage projects were hosted in historical spaces refurbished and renovated especially for the ECoC, such as the Barbara café, the Depot History Center, the White Stork Synagogue and the Centennial Hall. In terms of decentralization of cultural fare in the long-term, one of the successful projects was the opening of the Fama Library and Cultural Center in the Psie Pole outskirts district.

3.1.2. Local identity and sense of belonging

During the period leading up to the ECoC bid, the main narrative of multicultural identity was reaffirmed with such statements as 'Breslau-turned-Wrocław became a laboratory in which Polish culture imported from the Eastern Borderlands flourished on material heritage left behind by the Germans and has undergone a metamorphosis into a unique cultural formation' (Chmielewski 2011, 12). Discussing the city's multicultural past was the main reflection of its heritage found in the bid book, and a number of projects dealing with this problem were executed, such as the exhibition 'Pojednanie/Versöhnung in progress. The Catholic Church and Polish-German relations after 1945', focused on the difficult and still unfinished process of Polish-German reconciliation since WWII.

Heritage was also a topic of some initiatives organized within the Backyard Door project. Based on a co-creation process between artists and residents of the tenement

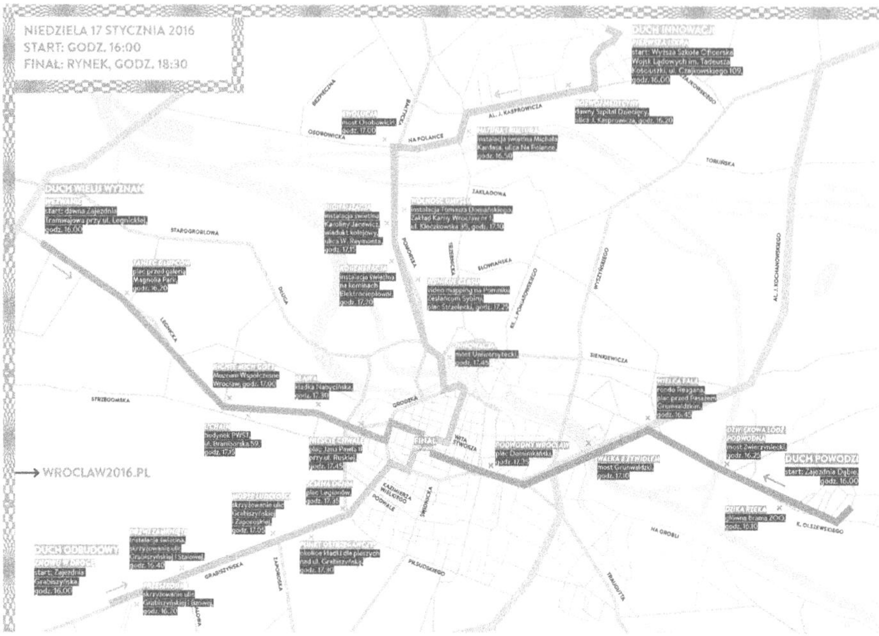

Figure 1. The routes of the Four Spirits Parade culminating in the Market Square during the Awakening Ceremony. Source: Wrocław 2016 webpage. Event leaflet. http://www.wroclaw2016.strefakultury.pl/weekend-otwarcia-esk-2016-startujemy

Figure 2. Locations of ECoC 2016 heritage-related projects in the historic center of Wrocław. Source: ICC, I. Morawska.

Figure 3. Localizations of ECoC 2016 heritage-related projects in Wrocław. Source: ICC, I. Morawska.

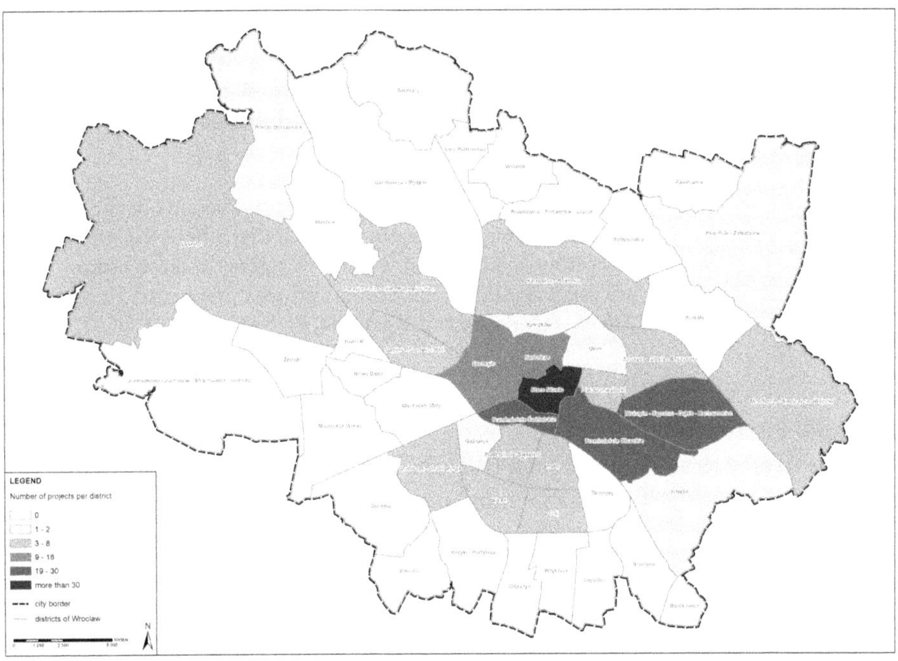

Figure 4. ECoC 2016 heritage-related projects per district in Wrocław. Source: ICC, I. Morawska.

houses in often neglected and socially excluded neighbourhoods of Ołbin, Gaj, Przedmieście Oławskie, Huby and Kozanów, the project aimed to activate the residents, commit them to their immediate surroundings, and consequently to strengthen their local identity and feeling of belonging. 61 artists took part in the project, and 32 projects were carried out in 42 backyards and seven housing estates. 70 meetings and consultations with residents took place. 4,400 participants took part in the activities – primarily residents of Wrocław (Zarzycki 2016, 176–177).

The project has been rather positively evaluated (Krysiński and Banaś 2017) and is still continued. One of the success factors was exploring and using the local content important for residents, such as the history of a particular place and positive memories to actively engage them in creating installation art that communicates what was meaningful and valuable for them (27). However, addressing difficult and unworked issues related to collective memory does not always serve social activation, which should be devoid of content that could lead to self-exclusion from activation activities. Referring to traumatic events that bring to mind negative emotions and memories may destroy the purpose of the intervention, which is, inter alia, getting to know the inhabitants, strengthening social ties and involving members of local communities in activities aimed at activating the development potential or eliminating social problems (37). Such a situation was observed in case of an unsettling work, found controversial and unintelligible, contributed by Artur Żmijewski. Decaying, forgotten old German tombstones (probably from a German cemetery turned into a park as part of the de-Germanization campaign of the 1940s) were arranged along a sidewalk to confront residents with the repressed past of the city and the blurring of its heritage.

Another project that was reported to contribute to local identity and sense of belonging was the microGRANTS. The evaluation report claims that local projects enabled residents to learn about the history of their immediate surroundings, which might influence the development of (yet non-existent) identity of the neighbourhoods. As declared by its participants,

> I think it is something (...) that might be starting. (...) I was participating in one of the excursions in Szczepin and it started, for example, [people recalling that] there was an open-air market, people remember what it looked like (...). I think it is great because history is not something that happened 100 years ago but is important also what happened 10 years ago, that suddenly disappeared. (Dolińska, Kajta, and Niedźwiecka-Iwańczak 2017b, 47)

3.1.3. Social interactions

If culture is treated as a relational good (Bianchi 2003, 395–397), participation in culture could be a way to deepen social interactions and as a result improve individual well-being (Blessi 2014). Moreover, the relationship between urban space and cultural participation is crucial for both socialization of individuals and building social capital. In Wrocław, 'by participating in culture in the non-private sphere – in the public sphere of the city, and in addition by not participating alone but sharing the experience with companions, a significant capital of aesthetic and emotional experiences is built' (Kajdanek 2017, 7). When describing individual benefits from taking part in ECoC projects, respondents emphasized that it was an opportunity to meet friends and family as well as a conversation piece with friends and colleagues (Pluta 2017, 24).

The integration effect of the microGRANTS projects was noticed both by the participants ('we heard numerous times that projects such as ours (…) provide a chance to break the neighborly ice and give a chance to meet people', 'many new neighborly acquaintances were made during this time') and the among the organizers of the projects themselves. Integration of the local culture sector and establishing new relations were reported: 'Such projects contribute to and provoke cooperation in local communities. They connect the local government (…), business, (…) and non-governmental (…) sectors' (Dolińska, Kajta, and Niedźwiecka-Iwańczak 2017b, 45). Also, the Backyard Door project succeeds in establishing or strengthening social ties, especially when the artistic activities were based on the involvement of residents in the designing and production process. It is well illustrated by a statement of one of the residents: 'Now we talk with each other, visit each other at home. When I go through the backyard children shout greetings. It is an amazing change' (Krysiński and Banaś 2017, 25).

3.2. Relations with the urban space and built heritage

The ECoC 2016 was meant as a tool to animate urban space by encouraging residents' activities (especially cultural) in the public space and mitigate the problem of 'public agoraphobia' diagnosed in the bid book. Wroclaw was the first ECoC that introduced architecture as one of the eight curatorial domains, therefore, giving tangible heritage additional importance in the project. Zbigniew Maćków, the curator, sees a number of longer-term effects of this move, including increasing appreciation for modernist architecture that started with renovation of the pre-war WUWA[5] (Strefa Kultury Wrocław 2021). It was in fact the WUWA estate that inspired the idea to create a contemporary model estate tailored to real people's needs – Nowe Żerniki.

The relationship residents have with the urban space, including built heritage, is affected by physical and practical features of the space that allow safety to walk, meet others and enjoy and express oneself (i.e. good scale and good atmosphere) as well aesthetic visual quality of the city (the way buildings and spaces are designed, the composition of details, and above all their mutual relations in the urban context). The composition of squares as meeting places, art in the urban space, greenery, the way of illuminating the city and its architecture at night create positive sensory experiences, which are one of the twelve criteria of good quality of urban spaces created by Jan Gehl (2010, 2011). His ideas were known to Wrocław Municipality (study visit, workshops and lectures, 2010–2011) and might have been considered while deciding on regeneration processes in the city. In the context of quality of life and ECoC, Wrocław residents' relations with the built environment can be presented from three different angles: revitalization of degraded neighbourhoods and venues, introducing cultural life into less obvious heritage spaces and upgrading public space.

3.2.1. Revitalization

The ECoC 2016 summary report lists the revitalization processes, which were part of a wider city strategy implemented by the municipal corporation Wrocławska Rewitalizacja Ltd., undertaken in Nadodrze and Psie Pole as well as the Four Denominations District. The latter is unique for its religious diversity manifested by the presence of houses of worship of several different religions and denominations: Catholicism, Judaism,

Protestantism and Eastern Orthodox. Among the results of the revitalization here, organized jointly by the clergy and NGOs, was the conservation of the White Stork Synagogue, which was frequently used as an ECoC venue. The Nadodrze and Psie Pole districts were gradually revitalized, using also the potential of local heritage and upgrading cultural life with the involvement of several NGOs.

Another type of ECoC infrastructural project was renovation and adaptation for cultural functions. It includes work in postindustrial heritage spaces, such as Ruska 46, Krzywy Komin, and the Depot History Centre, as well as a former air-raid shelter from 1942 turned into Wrocław Contemporary Museum. The Pan Tadeusz Museum is housed in a beautiful converted tenement house, Under the Golden Sun on the Market Square, as a foil for the only existing manuscript of Adam Mickiewicz's *Pan Tadeusz* (1834), the national epic of Poland. This is a symbolic liaison with Lviv, as the Ossolineum collection was moved from that city to Wrocław after WWII. The Capitol Music Theater, a pre-war cinema, reopened to the public with extended space, its valuable interiors renovated and new technology installed for improved cultural experiences. The Na Grobli Studio of the Grotowski Institute was moved into the renewed and modernized Wratislavia Rowing Association building, most of whose Art Nouveau furnishings and fittings are still visible. The Henryk Tomaszewski Theater Museum is housed in what is left of the southern wing of the Royal Palace of the kings of Prussia. And, last but not least, Barbara, a café which enjoyed cult status for its original interior and history of anti-communist protests, was refurbished to serve as the ECoC information center and is now used as the Culture Zone Wrocław (the ECoC legacy agency) headquarters.

3.2.2. Cultural projects in non-obvious heritage spaces

The need for new cultural spaces provided an important impetus for popularization of heritage venues that had not previously been directly associated with cultural activity. The Forgotten City saw open-entry concerts performed on twelve extraordinary stages: pre-war tenement courtyards, historic squares, interiors of Art Nouveau houses, historic railway stations, a former psychiatric and nervous system disease clinic and postindustrial spaces. Another example is the use of the Oder riverbanks and bridges in the Bridge Builders and Flow projects. In Bridge Builders, the discussion on the complexity of the city's history focused on Wrocław's bridges – the majority of which are of historical and artistic value. The idea behind the event was to offer small financial grants and methodological support to local activists, whose work was to focus, in terms of both content and location, on one of the bridges. The focal point of Flow was the Oder River, which, together with the surrounding parks, buildings and bridges, was the stage for various performances and cultural meetings on the city's history done by local and European artists. Offering cultural events in non-obvious, historical venues created an opportunity for people to appreciate and rediscover heritage as a part of their everyday lives. It focused public attention on their surroundings, unveiling their history and artistic value even in forgotten parts of the city.

3.2.3. Upgrading heritage-related public space

A project that stands out from all the others in upgrading public space is a relatively small-scale initiative, 'Spaces for Beauty. Urban Public Space' carried out by the

Municipal Art Consultant, whose achievement was the restoration of 14 run-down art installations and sculptures in urban public and introduction of new art into the public space. It was based on a conviction that art in public space is a way to improve the quality of the space, emphasize its significance and improve its visual attractiveness. Furthermore, the ECoC boosted awareness of the city's beauty and promoted the rediscovery of the value of historical architecture.

> The things that went on in the context of heritage – in terms of promotion and exhibitions, dissemination, education – directly helped this sort of general awareness of these monuments and their value to such a degree that it is absolutely a direct impact [of the ECoC]. (WroCons01)

4. Discussion

The above sections focused on these heritage-related ECoC 2016 projects that can impact quality of life via access to and participation in culture as well as relations with the urban space and built heritage. The main points of the discussion that follows are summarized in Table 2.

In Wrocław 2016, access to culture and active participation was explicitly regarded as 'a method for reconnecting the excluded to the shared world' (Chmielewski 2011, 10) so as to increase their quality of life and a way 'to build the stock of public trust necessary to strengthen our social development' (7). The 'excluded' category encompassed both those excluded due to their social and/or economic situation as well the self-excluded (resulting from the declared lack of time, interest, awareness, etc.). No study has been performed to directly verify the degree of successful implementation of these goals. However, the number of participants in the events and evaluation reports of specific projects suggest that participation in the ECoC cultural activities could have contributed to the increase in quality of life.

First of all, the microGRANTS scheme overcame a passive and indifferent attitude of participants towards engagement in the cultural life of the neighbourhood, increasing interest in local history and heritage (Dolińska, Kajta, and Niedźwiecka-Iwańczak 2017a). Second, the ECoC cultural program offered a diverse spectrum of possibilities for entertainment and learning. Among the most frequently quoted individual benefits from participating in the ECoC projects, respondents listed 'an interesting way to spend my free time', 'an opportunity to have fun', 'arousing positive emotions', 'an opportunity to learn/experience something new or unique', 'expanding knowledge' and 'stimulating reflection and independent interpretations' (Pluta 2017, 24). On the basis of the literature review, we can assume that such assessments of participation in the ECoC suggest a growth of individual satisfaction and self-development (Di Giovinazzo 2011, 20). Blessi (2014) also emphasize a positive connection between culture as a relational good and individual well-being (feeling of community; increasing one's capabilities). Interestingly, there is also some literature that reflects on the positive correlation between cultural participation and subjective evaluation of one's health (e.g. Grossi 2012; Leadbetter and O'Connor 2013; Hill 2013). It should be noted, however, that Pluta et al. are quite cautious when discussing long-term effects on cultural consumption patterns. They speak of 'the moderate effectiveness of the supply impulse' of the ECOC program

Table 2. The potential of ECoC 2016 to contribute to quality of life.

Contributors to quality of life	ECoC 2016 heritage-related projects	Potential impact on quality of life
Access to and participation in culture	• Backyard Door project • microGRANTS scheme • Flow Quartet • ECoC parks	• Entertainment and learning • Individual satisfaction and self-development • Subjective evaluation of one's health • Local identity • Sense of belonging • Socialization of individuals • Building social capital
Relations with the urban space and built heritage	• The Forgotten City concerts • Bridge Builders and Flow • Spaces for Beauty. Urban Public Space • Revitalization projects in the Nadodrze, Psie Pole, and Four Denominations districts • New cultural infrastructure, venues, and institutions	• Local identity • Sense of belonging • Sense of security • Aesthetic pleasure • Entertainment and learning • Socialization of individuals • Building social capital

Source: own elaboration.

and the consumption 'strongly influenced by status factors, differentiating the social position of respondents and their socio-cultural profiles, which influenced both the segmentation of 'cultural consumers' and made it possible to easily disclose circuits of culture' (2017, 36–37).

One way to improve the level of cultural participation is to bring activities closer to residents. At least 41% of the residents (Pluta 2017, 10) noticed the process of bringing culture closer to people by organizing cultural projects in venues outside traditional cultural infrastructure, including outdoors – parks, streets and squares. There was also an attempt to disperse ECoC activity throughout the city. However, a general feeling was that 'the ECoC celebrations failed to overcome the fears of doing something outside the center' (Banaszak 2017b, 21), and especially heritage-related projects were implemented mostly in the historic city center.

Heritage exists in close relation to identity (which in turn plays a vital role in quality of life), being its symbolic form where the social sense of belonging is built and the community's symbolic scope is delimited (Kowalski 2013, 98). History is also used as an important tool to shape the socio-cultural identity of a place and its brand (Ashworth 1994, 13). With heritage being an immanent and often deeply personal part of individuals' lives, projects that deal with it have the power to raise attention and social engagement (either to protect and cherish it or to contradict and discuss).

It is worth noticing, then, that cultural mega-events are frequently used as opportunities for reflection on painful collective memories, political reconciliation processes, or controversial moments from a city's past (Bianchini and Borchi 2018, 197). The guiding concept of Wrocław's bid for the ECoC title was 'Metamorphoses of Cultures', a reference to the city's complex multicultural past. Wrocław has always been a city in transition – due not only to the frequency of wars and conflicts that have affected it, but also to its shifting state affiliation (Czech, Polish, Habsburg (Austrian), Prussian and after WWII Polish again) which brought repeated changes to its legal, economic and cultural systems and values. It must be noted that Wrocław is the largest city in the history of mankind in which there was almost a complete population exchange (after WWII). Those German residents who had not been evacuated during the siege of Festung Breslau were forced to move out, being replaced by Polish settlers, mostly from Lvov (now Lviv, Ukraine), cities lost by Poland to the Soviet Union and other parts of Poland. For them, their new home was imbued almost entirely by the heritage of the enemy. That proved to be a greater challenge for the city than the reconstruction of its material fabric, 70% of which (counting the historic center) had been destroyed or damaged by the war. For almost 50 years the official line was to find evidence of the Polishness of the city while passing over or downplaying the hundreds of years of other influence. It was only after 1989 that not only the political and economic system changed but also the approach to the city's past.

The multicultural past of the city was an important backbone of the ECoC, with a number of projects undertaking this issue, including the Backyard Door, microGRANTS and the Flow Quartet. Although no evaluation of their impact on the local identity and the perception of the past was done, it seems that there could be no coincidence that a survey performed right after the ECoC year showed a rise by 15.3 percentage points in the group of respondents that chose 'a microcosm where influences of different cultures have mixed over centuries' as the best way to describe Wrocław (Dolińska and Makaro 2013; Makaro and Dolińska 2017) – see Table 3.

Quality of life can also be influenced by the quality of public space (Weichert 1984, 21), including the presence of cultural heritage. Paszkowski even claims that cities with historical urban fabric, diverse architecture, public space and a small scale are perceived by residents to be much better than modern but impersonal neighbourhoods (2011, 227). The level of an individual's sense of security and the degree to which they establish contacts with the surrounding world depends on how the urban space is organized, because this is a point of reference, organizes human activities and reflects the knowledge and values of a given community (Weichert 1984, 4). People strive to tame the space around them by giving it functional, aesthetic and cultural order so as to make the world around them more friendly, understandable and familiar (Karwińska 1998, 53).

Pluta emphasizes that the ECoC in Wrocław was deeply rooted in the public space of the city (Banaszak 2017a, 52). It influenced not only high rates of participation in specific projects but also, one can argue, 'taming' or 'domestication' of the public space as a place. With that in mind, I argue that organizing cultural events in non-obvious heritage venues also contributed to familiarization of the space around them and, in consequence, potentially also a raise in the feeling of being more safe in neighborhoods where venues were located (in many cases, it was discovering participants' own neighbourhoods). Similarly, revitalization of neglected areas can also be seen as a contributor to the potential rise in quality of life. ECOC's new cultural infrastructure and the institutions it comprises may furthermore provide an opportunity for greater access to culture, a varied cultural offer and a new 'third space', as well as employment and volunteering opportunities.

ECoC 2016 seems to have contributed to the increase of residents' awareness and appreciation of the surrounding heritage, as well as to identity building and feeling of belonging. One representative of the municipal authorities states that the Spaces for Beauty project (along with the Spaces for Beauty concept as a whole) increased public support for spending public money on aesthetic issues, something that had not been so obvious before 2016 (WroMun02). This could be interpreted as a higher awareness

Table 3. Main characteristics of Wrocław according to survey results from 2011 and 2017.

Characteristics of Wrocław	Survey results	
	2011	2017
A microcosm where influences of different cultures have mixed over the centuries	46%	61.3%
Since WWII a Polish city dating back to Piast times	29%	22.8%
Since WWII part of 'the Eastern borderlands', as many settlers came to Wrocław from there	15%	8%
Most of all a German city, because the Germans were the ones that created the city almost just as we know it today	10%	1.9%

Source: own based on Dolińska and Makaro (2013, 77) and Makaro and Dolińska (2017, 145–146).

and appreciation of the aesthetics of public space (also listed as a contributor to quality of life). Another expert, long involved in the memory and heritage issues in Wrocław, sees some ECoC influence when claiming that 'people start to be interested in what kind of house they live, what its history is, who lived there before. When I organize a guided walk, 500–600 people [sign up]!'. Moreover, the attitude toward historical buildings has evolved. There was a program 'spun up by the ECoC, used as a tool, a so-called tenement building program of co-financing even 70% renovation costs of listed buildings'. At first, residents were quite suspicious and reluctant to use it. But then, 'a rivalry between neighbors began. Something absolutely fantastic started – people started to apply for the listing themselves (...), they are proud their building is listed, that it raises its value'. The same expert argues that the residents have started to value their monuments more. In the past, a dozen desperate people, 'mainly renowned professors', protested demolitions of historic buildings. 'And now, when they want to demolish something, people immediately raise their alarm that they will not let it in their lives' (WybPap01).

It is also worth mentioning that, according to one of the experts, the Spaces for Beauty concept of renovations and revitalizations definitely worked. In her opinion, they should not be treated as 'an ordinary renovation – it is something that also changes awareness. You have to remember that the change of space and surroundings makes people think differently. They behave differently – that's what pulls them up' (WybPap01). Targeting the aesthetics and friendliness of public space was an aspect of the Spaces for Beauty concept that not only contributed to quality of life, but can also be seen as a way to foster competitiveness and attractiveness to both residents and the outside world. It is also an angle from which to view the Reclaiming Beauty and New Spaces for Beauty ECoC themes.

5. Conclusions

Quality of life depends on various factors, and the research discussed above argues that cultural heritage can play a role in its improvement. Wrocław 2016, a cultural mega-event comprising 425 projects and 2000 events (10% categorized as heritage-related), was a good opportunity to mobilize heritage, both tangible and intangible, to support improvements in quality of life. In fact, it can be viewed as a time of connecting heritage mobilization and harnessing its pro-development potential. Based on the qualitative desk research, I argue that projects discussed, especially microGRANTS, Backyard Door and regeneration and renovation initiatives have likely triggered changes in quality of life – including access to culture, fostering local integration, pride and identity and improved quality of public space and infrastructure. More in-depth research, however, is needed to analyse the long-term effects. With any projects, but especially those as large as ECoCs, the best way to verify potential impact is to make a baseline report before the whole process starts, to be able to actually measure any changes occurring in its wake. Unfortunately, Wrocław 2016 did not conduct any baseline research. This could have provided more evidence to examine the relation between heritage and quality of life, and support arguments raised in this article. Such research is important not only from the academic but also policy-making perspective. It allows firstly, to broaden the variety of tools at a city's disposal (heritage and its socio-economic potential still being underestimated) and, secondly, to examine the effectiveness of actions implemented to achieve the city's strategic goals (here, in particular, those related to quality of life).

The legacy of Wrocław 2016 includes new cultural infrastructure with the potential to support the development of social capital and the continuation of some ECoC projects, the common experience of Wrocław residents (contributing to their identity, pride and feeling of belonging and cultural engagement), increased potential of the culture sector, international recognizability and a new municipal strategy for culture (Banaszak 2017a, 2). Importantly, there seems to be a change in attitudes towards cultural heritage. As one interviewee claimed:

> [ECoC] has shown very clearly that monuments and heritage, historical objects and so on, that they are the real value in this city, one which translates directly into money, tourist traffic; they are the actual attractors. And when we work with them and put our faith in them, there will be tangible benefits. (WroCons01)

One could claim that similar positive results can be achieved without a mega-event that drains the public budget with festivities and away from real problems and daily cultural life. While it is true that the ECoC budget seems enormous (see Table 4), and 78% came from public sources, it should be emphasized that Wrocław 2016 was an all-encompassing event with a number of projects either aligned with or under the umbrella of ECOC 2016; consequently, there were few cultural projects that could not obtain public financing due to the ECoC focus. In fact, the ECoC was on one hand a structural framework for inspiring and motivating actors and entities to develop new projects and ideas, often rooted in local communities and meant for broadening social participation; on the other, it was a pretext for applying for additional financial resources from outside the City of Wrocław. The latter was especially true in the case of infrastructural projects where the ECoC presumably played the role of an additional benefit or value for the panel

Table 4. ECoC 2016 sources of financing OPEX costs.

Sources of financing	Income in the years 2011–2016 (EUR)	%
City of Wrocław	32,699,843.07	39
Other public entities (incl. Ministry of Culture National Heritage and the Multiannual Program European Capital of Culture 2016)	30,715,916.12	37
EU funds	1,793,475.68	2
Sale (e.g. tickets, publications)	10,198,535.36	12
Sponsors and barters	6,641,780.40	8
Other	1,689,809.31	2
Total	83,739,359.93	100

Source: Banaszak (2017a, 169). Own conversion of PLN to EUR based on the average exchange rate of the European Central Bank for 2011–2016 (EUR 1 = PLN 4.2058).

evaluating applications for the European Structural Funds or the EEA Financial Mechanisms (e.g. the Depot History Center or Four Dome Pavilion). Arguably, this alignment and ECoC labelling resulted in a final program not abstracted from the specific Wrocław context, engaging stakeholders and partners to implement their own ideas, strengthening relationships and providing additional resources to many important initiatives for the city, a number of which can be associated with improvements in quality of life, as analysed in this article.

Notes

1. With no special category for heritage projects in ECoC 2016, it is even difficult to assess the part of ECoC's budget spent on heritage. Among the influences ECoC 2016 had on heritage one could list: an increase in the amounts spent on cultural infrastructure, renovation and revitalization, and a change in residents' attitude towards heritage.
2. The main indicator that a given project addressed heritage was that it dealt with identity, memory, and history on the one hand, and with renovation or regeneration on the other.
3. The hypothesis is that the collocation most used with 'heritage' is 'national', which is complicated in the case of Wrocław. Moreover, the city, as claimed, is more forward-looking than dwelling on its past (FGI02; Chmielewski 2011, 6).
4. This number includes both residents and visitors. There is no exact data concerning the number of the former, but the research on cultural participation cited in the article refers to the residents of Wrocław and the region. To compare, Wrocław is a city of 642 869 inhabitants (as of 31.12.2019, Statistics Poland, https://bdl.stat.gov.pl/BDL/dane/podgrup/tablica)
5. In 1929, Wrocław hosted a Living and Work Space Exhibition (Wohnung und Werkraum Ausstellung, WuWA) presenting model housing estates created by the Deutscher Werkbund. Buildings created for the exhibition were meant to be just prototypes for mass housing estates, but most of them survived WWII and have since been inhabited.

Disclosure statement

No potential conflict of interest was reported by the author(s).

Funding

Sections of this paper are based on and supported by the research conducted within the JPI Cultural Heritage research project 'HOMEE – Heritage Opportunities/Threats within Mega-Events in Europe', with a grant issued by the Minister of Culture and National Heritage of the Republic of Poland.

ORCID

Joanna Sanetra-Szeliga http://orcid.org/0000-0002-8679-2074

References

Ashworth, G. 1994. "From History to Heritage, from Heritage to Identity: In Search of Concepts and Models." In *Building a New Heritage: Tourism, Culture and Identity in the New Europe*, edited by G. J. Ashworth and P. J. Larkham, 13–30. London: Routledge.

Banaszak, E., J. Banaś, E. Bendyk, M. Błaszczyk, K. Dolińska, R. Drozdowski, and B. Gierat--Bieroń, et al. 2017a. *Spaces for Beauty: Summarized.* Wrocław: Uniwersytet Wrocławski, Urząd Miejski Wrocławia, Biuro Festiwalowe IMPART 2016.

Banaszak, E., M. Błaszczyk, K. Kajdanek, and J. Pluta. 2017b. *Terra (in)cognita. Europejska Stolica Kultury w perspektywie wrocławskich instytucji kultury. Analiza formalna.* Wrocław: Uniwersytet Wrocławski, Urząd Miejski Wrocławia, Biuro Festiwalowe IMPART 2016.

Bianchi, M. 2003. "A Questioning Economist: Tibor Scitovsky's Attempt to Bring Joy Into Economics." *Journal of Economic Psychology* 24. doi:10.1016/S0167-4870(03)00019-9

Bianchini, F., and A. Borchi. 2018. Heritage in the European Cities and Capitals of Culture Programmes. Reflections on Some Key Themes and Issues. *Cartaditalia* I (Special Issue).

Blessi, G. T., E. Grossi, P. L. Sacco, G. Pieretti, and G. Gerilli. 2014. "Cultural Participation, Relational Goods and Individual Subjective Well-Being: Some Empirical Evidence." *Review of Economics & Finance* 4 (3): 33–46.

cg. 2012. Od dużego Expo po World Games - o co starał się Wrocław. *sport.pl.* http://www.wroclaw.sport.pl/sport-wroclaw/56,123436,10981452,Od_duzego_Expo_po_World_Games___o_co_staral_sie_Wroclaw.html

Chmielewski, A., M. A. Gerlich, M. Jasiński, A. Jerska, A. Kołodyńska, A. Krasnowska, and E. Łyczkowska, et al. 2011. *Spaces for Beauty. Revisited. Wrocław's Application for the Title of European Capital of Culture 2016.* Wrocław.

Constanza, R., B. Fisher, S. Ali, C. Beer, L. Bond, R. Boumans, and N. L. Danigelis, et al. 2008. "An Integrative Approach to Quality of Life Measurement, Research, and Policy." *S.A.P.I.EN.S* 1 (1). http://journals.openedition.org/sapiens/169

Davies, J., R. Whimster, and L. Clayton, eds. 2009. *Heritage Counts 2009.* English Heritage. https://historicengland.org.uk/content/heritage-counts/pub/HC09_England_Acc-pdf

Deloitte. 2012. *Raport: Podsumowanie kosztów i oszacowanie korzyści z organizacji turnieju UEFA EURO 2012.* Accessed June 15, 2015. www.deloitte.com/view/pl_PL/pl/branze/sport/24b45edccc8db310VgnVCM30000003456f70aRCRD.htm ()

Di Giovinazzo, V. 2011. *Consumer Preferences and Individual Satisfaction. Tibor Scitovsky on the Role of the Arts in Society.* HES (History of Economics Society), University of Notre Dame. Accessed August 2, 2016. https://econ.duke.edu/uploads/media_items/consumer-preferences-and-individual-satisfaction-tibor-scitovsky-on-the-role-of-the-arts-in-society.original.pdf

Dolińska, K., J. Kajta, and N. Niedźwiecka-Iwańczak. 2017a. "'Mieszkańcy dla mieszkańców'. Partycypacyjny charakter Programu „mikroGRANTY ESK 2016"." In *Raport „mikroGRANTY ESK 2016" Edycja 2016*, edited by K. Dolińska, 10–41. Wrocław: Uniwersytet Wrocławski, Urząd Miasta Wrocławia, Biuro Festiwalowe IMPART 2016.

Dolińska, K., J. Kajta, and N. Niedźwiecka-Iwańczak. 2017b. "Ocena wydarzeń realizowanych w ramach Programu "mikroGRANTY ESK 2016"." In *Raport „mikroGRANTY ESK 2016" Edycja 2016*, edited by K. Dolińska. Wrocław: Uniwersytet Wrocławski, Urząd Miasta Wrocławia, Biuro Festiwalowe IMPART 2016.

Dolińska, K., and J. Makaro. 2013. *O wielokulturowości monokulturowego Wrocławia.* Wrocław: Wyd. Uniwersytetu Wrocławskiego.

Filip, M. 2012. *Economic Impacts of UEFA Euro 2012 Staging in Wroclaw.* MPRA Paper No. 58207. https://mpra.ub.uni-muenchen.de/58207/

Fujiwara, D., L. Kudrna, and P. Dolan. 2014. *Quantifying and Valuing the Wellbeing Impacts of Culture and Sport.* Department for Culture, Media & Sport. https://assets.publishing.service.gov.uk/government/uploads/system/uploads/attachment_data/file/304899/Quantifying_and_valuing_the_wellbeing_impacts_of_sport_and_culture.pdf

Galar, R., et al. 2006. *The Strategy "Wroclaw in the Perspective 2020 Plus".* https://bip.um.wroc.pl/uploads/files/ProgramMiejski/strategia_Wroclaw2020plus_en.pdf

García, B. 2004. "Urban Regeneration, Arts Programming and Major Events. Glasgow 1990, Sydney 2000 and Barcelona 2004." *International Journal of Cultural Policy* 10 (1). doi:10.1080/1028663042000212355

García, B., and T. Cox. 2013. *European Capitals of Culture: Success Strategies and Long-Term Effects Study.* Brussels: Directorate-General for Internal Policies. Policy Department. Structural and Cohesion Policies, European Union.

Gehl, J. 2010. *Cities for People.* Washington/Covelo/London: Island Press.

Gehl, J. 2011. *Life Between the Buildings.* Washington/Covelo/London: Island Press.

Giovanangeli, A. 2015. "Marseille, European Capital of Culture 2013 Ins and Offs: A Case for Rethinking the Effects of Large-Scale Cultural Initiatives." *French Cultural Studies* 26 (3). doi:10.1177/0957155815587236

Giraud-Labalte, C., K. Pugh, Š. Quaedvlieg-Mihailović, J. Sanetra-Szeliga, B. Smith, A. Vandesande, and C. Thys. 2015. *Cultural Heritage Counts for Europe. Full Report.* Kraków: International Cultural Centre.

Gómez, M. 1998. "Reflective Images: The Case of Urban Regeneration in Glasgow and Bilbao." *International Journal of Urban and Regional Research* 22 (1). doi:10.1111/1468-2427.00126

Grossi, E. 2012. "The Interaction Between Culture, Health and Psychological Well-Being: Data Mining from the Italian Culture and Well-Being Project." *Journal of Happiness Studies* 13 (1). doi:10.1007/s10902-011-9254-x

Hall, P. 2014. *Good Cities, Better Lives. How Europe Discovered the Lost Art of Urbanism.* London/New York: Routledge.

Hermant-de Callataÿ, C., and C. Svanfeldt. 2011. *Cities of Tomorrow. Challenges, Visions, Ways Forward.* Luxembourg: European Commission, Directorate General for Regional Policy, Publications Office of the European Union.

Herrero, L. C., J. Á. Sanz, M. Devesa, A. Bedate, and M. J. del Barrio. 2006. "The Economic Impact of Cultural Events: A Case-Study of Salamanca 2002, European Capital of Culture." *European Urban and Regional Studies* 13 (1). doi:10.1177/0969776406058946

Hill, K. 2013. "The Arts and Individual Well-Being in Canada: Connections Between Cultural Activities and Health, Volunteering, Satisfaction with Life, and Other Social Indicators in 2010." *Statistical Insights on the Arts January* 11 (2). https://hillstrategies.com/wp-content/uploads/2013/01/Arts_well_being2010.pdf

Jones, Z. 2020. *Cultural Mega-Events. Opportunities and Risks for Heritage Cities.* Abingdon/Oxon/New York: Routledge.

Kajdanek, K., E. Banaszak, M. Błaszczyk, and J. Pluta. 2017. *Specjalne Weekendy ESK. Raport Podsumowujący.* Wrocław: Uniwersytet Wrocławski, Urząd Miejski Wrocławia, Biuro Festiwalowe IMPART 2016.

Karwińska, A. 1998. *Wartości i potrzeby społeczne a przemiany środowiska miejskiego.* Kraków: Wydawnictwo Akademii Ekonomicznej w Krakowie.

Kowalski, K. 2013. *O istocie dziedzictwa europejskiego – rozważania.* Kraków: Międzynarodowe Centrum Kultury.

Krysiński, D., and J. Banaś. 2017. *Wrocław – wejście od podwórza. Raport ewaluacyjny projektu.* Wrocław: Uniwersytet Wrocławski, Urząd Miasta Wrocławia, Biuro Festiwalowe IMPART 2016.

Langen, F., and B. García. 2009. *Measuring the Impacts of Large Scale Cultural Events: A Literature Review. Impacts 08.* Liverpool: University of Liverpool, Liverpool John Moores University. https://www.liverpool.ac.uk/media/livacuk/impacts08/pdf/pdf/Impacts08-FLangen_and_BGarcia_May_2009_Events_Review.pdf

Lähdesmäki, T. 2014. "European Capital of Culture Designation as an Initiator of Urban Transformation in the Post-Socialist Countries." *European Planning Studies* 22 (3). doi:10.1080/09654313.2012.752438

Leadbetter, C., and N. O'Connor. 2013. *Healthy Attendance? The Impact of Cultural Engagement and Sports Participation on Health and Satisfaction With Life In Scotland.* Scottish Government Social Research. https://www.gov.scot/publications/healthy-attendance-impact-cultural-engagement-sports-participation-health-satisfaction-life-scotland/documents/

Makaro, J., and K. Dolińska. 2017. "Postrzeganie Zróżnicowania Etnicznego, Stosunek do „Innych" i Poczucie Bezpieczeństwa." In *Wrocławska Diagnoza Społeczna 2017. Raport z Badań Socjologicznych nad Mieszkańcami Miasta,* edited by K. Kajdanek and J. Pluta.

Wrocław: Uniwersytet Wrocławski. http://www.repozytorium.uni.wroc.pl/Content/89133/WROC%C5%81AWSKA%20DIAGNOZA%20SPO%C5%81ECZNA%202017%20RAPORT.pdf
Matarasso, F. 1997. *Use Or Ornament? The Social Impact of Participation in the Arts*. Stroud: Comedia.
Müller, M., and J. Pickles. 2015. "Global Games, Local Rules: Mega-Events in the Post-Socialist World." *European Urban and Regional Studies* 22 (2): 121–127. doi:10.1177/0969776414560866
Ni, P. 2012. *The Global Urban Competitiveness Report 2011*. Northampton: Cheltenham.
Oldenburg, R. 1989. *The Great Good Place: Cafes, Coffee Shops, Community Center, Beauty Parlors, General Stores, Bars, Hangouts, and How They Get You Through the Day*. New York: Paragon House.
Paris, D., and T. Baert. 2011. "Lille 2004 and the Role of Culture in the Regeneration of Lille Métropole." *TPR* 82 (1): 29–43. doi:10.3828/tpr.2011.7
Paszkowski, Z. 2011. *Miasto idealne w perspektywie europejskiej i jego związki z urbanistyką współczesną*. Kraków: Universitas.
Pluta, J., E. Banaszak, M. Błaszczyk, and K. Kajdanek. 2017. *Uczestnictwo w kulturze w perspektywie Europejskiej Stolicy Kultury Wrocław 2016. Raport z badań CATI z udziałem mieszkańców Wrocławia i województwa dolnośląskiego*. Wrocław: Uniwersytet Wrocławski, Urząd Miasta Wrocławia, Biuro Festiwalowe IMPART 2016.
Ponzini, D., Z. M. Jones, F. Bianchini, E. Tommarchi, J. Georgi, E. Dova, and A. Sivitanidou, et al. 2019. *HOMEE Literature Review of Mega-events. Addressing Cultural Heritage Issues*. www.tau-lab.polimi.it/homee-literature-review-regarding-mega-events-cultural-heritage
Psatha, E., A. Deffner, and Y. Psycharis. 2011. Defining the Quality of Urban Life: Which Factors Should Be Considered? *The 51st ERSA Congress, New Challenges for European Regions and Urban Areas in a Globalised World*. Barcelona: ERSA.
Richards, G. 2014. "Eventfulness and the Quality of Life." *Tourism Today*. Fall.
Richards, G., and R. Palmer. 2010. *Eventful Cities: Cultural Management and Urban Revitalisation*. Oxford: Elsevier.
Rogerson, R. 1999. "Quality of Life and City Competitiveness." *Urban Studies* 36. doi:10.1080/0042098993303
Sanetra-Szeliga, J. 2016. "Kultura i rozwój miasta. Wyzwania XXI Wieku." In *Pamięć, Wybór, Tożsamość. Szkice o Mieście*, edited by K. Broński, 183–204. Kraków: International Cultural Centre.
Sanetra-Szeliga, J. 2017. "Kultura jako element składowy jakości życia w mieście." *Kultura i Rozwój* 1 (2): 46–59. doi:10.7366/KIR.2017.1.2.03
Sanetra-Szeliga, J., and K. Jagodzińska. 2017. "The Socio-Economic Impact of Cultural Heritage. Theoretical Aspects." In *The Power of Heritage. Socio-Economic Examples from Central Europe*, edited by J. Sanetra-Szeliga and K. Jagodzińska. Kraków: International Cultural Centre.
Smith, A. 2012. *Events And Urban Regeneration. The Strategic Use of Events to Revitalise Cities*. Oxon/New York: Routledge.
Stone, P., and R. Sharpley. 2011. "Socio-cultural Impacts of Events: Meanings, Authorized Transgression and Social Capital (Chapter 23)." In *The Routledge Handbook of Events*, edited by S. Page and J. Connell, 347–361. London: Routledge.
Strefa Kultury Wrocław. 2021. *Kultura jest w nas! 5 lat temu wystartowała Europejska Stolica Kultury Wrocław 2016*. https://strefakultury.pl/news/kultura-jest-w-nas-5-lat-temu-wystartowala-europejska-stolica-kultury-wroclaw-2016
Weichert, K. 1984. *Elementy kompozycji urbanistycznej*. Warszawa: Wydawnictwo Arkady.
Zarzycki, J. 2016. *Wrocław – Backyard Door. An Overview*. Wrocław: Biuro Festiwalowe Wrocław. https://issuu.com/Wrocław2016/docs/wwop_2016_10_01
Interviews quoted
WroCons01. Wrocław Municipality Monument Conservator's Office. Wrocław 5/06/2019.
WroMun01. Wrocław Municipality. Kraków 26/03/2019.
WroMun02. Wrocław Municipality. Wrocław 5/06/2019.
FGI02. Focus group interview with ECOC Team. Wrocław 5/05/2019.
WybPap01. *Wyborcza* Daily. Wroclaw 19/06/2019.

OPEN ACCESS

Liverpool's European Capital of Culture legacy narrative: a selective heritage?

Tamara West

ABSTRACT

Liverpool hosted the European Capital of Cultural (ECoC) in 2008, four years after the city was granted UNESCO World Heritage Site status. A decade later the city council celebrated the anniversary of hosting, while at the same time UNESCO debated the imminent delisting of the city's world heritage site. Concentrating on data collected during this anniversary year, this paper critically examines the extent to which an ECoC legacy narrative in the city remains intrinsically linked to an economic regeneration-dominated understanding of the role of culture that has changed very little since the bidding and hosting process. Large scale events and visitor economy strategies dominate the cultural offer, whilst heritage, in particular the World Heritage Site, is at best drawn upon in property or tourism led redevelopment advertisement, or, at worst, seen as negatively competing with the progress of a perceived city 'renaissance'. This article draws upon cross sector multi-stakeholder mapping workshops, interviews and document analysis to explore how a selective legacy narrative continues to affect the development of a more embedded approach to cultural heritage within the city.

1. Introduction

In 2018 Liverpool City Council (LCC) celebrated the ten year anniversary of Liverpool 08, the hosting of the European Capital of Culture (ECoC). Described by one community stakeholder as 'all about making another big splash',[1] the Liverpool 18 anniversary events and associated statements sought to focus attention on Liverpool as a destination, a place that had, in the words of the Council's culture department Culture Liverpool 'got its swagger back' and, by 2018, the emphasis remained on 'putting us quite rightly as a cultural exemplar on a national and international stage'.[2] That same year, the UNESCO World Heritage annual meeting met to discuss the proposal for the imminent delisting of Liverpool's World Heritage Site. Inscribed in 2004, by 2012 it was on the world heritage in danger list and facing delisting by 2017. It is the environment that enabled these two quite different, but in several ways linked, events in 2017–2018 that this article explores in order to critically assess the extent to which the rhetoric of one negatively impacted on the realities of another.

This paper draws upon case study specific and regionally focused in-depth qualitative data collected precisely during the time of the ten year anniversary (2017–2018). As this coincided with the proposed de-listing of the UNESCO WHS and brought to the fore differences in the way the city is (re)presented, it provides a lens through which to explore the city's official legacy narrative of ECoC and set this against multi-stakeholder perspectives. Furthermore, this paper seeks to explore how the rhetoric of a success legacy continues to impact on the development of both culture and heritage within the city and the city region. Here the paper critically explores the enduring faith in culture as a driver for economic regeneration. It investigates how this has played a contributing role in the endangerment of the WHS and has hampered the understanding of wider heritage sites and practices in the city beyond the economic or touristic. To do this, it examines data from regional cross sector stakeholder mapping workshops alongside with interviews, document analysis and participant observations.

Firstly, the paper sets out the methodology utilized, including details of the mapping workshops and of the stakeholders and sectors involved. This is followed by an overview of the ECoC award and the context of Liverpool and the city's WHS via relevant literature. The article then discusses findings relating to the understanding, environment and challenges for culture and heritage in the city. Finally, the paper discusses the impact of what is here argued to be a highly selective legacy narrative. The paper contends that the use of evaluation and development of a selective success narrative can impact cultural and heritage policies. Economic strategies in the city still fail to engage fully with the idea of embedded or intrinsic cultural offer, preferencing a perceived need to buy in or add cultural value, and this approach is inherently linked to the city's hosting of ECoC. Here the paper suggests that a prevailing narrative of Liverpool 08 can also be seen to be reflected in both policy and academic discourse which, be it either from an overtly positive or a more critical viewpoint, has discussed the Liverpool ECoC as an almost mythical case in itself, devoid of wider context, and with this has largely failed to explore the impact of this on heritage practices and debates.

2. Definitions and methodology

This section introduces the definitions of ECoC, cultural heritage and governance that are used in the article. It provides details regarding the research methods as well. The data was gathered as part of a thematic project exploring shifts in cultural governance and investigated any shifts in the role and use of culture in related and unrelated policies and strategies in the city since the hosting of Liverpool 08. Whilst the data collected crosses several areas, this paper concentrates on the findings that relate to cultural heritage, economic development and ECoC legacy.

The ECoC programme, an annual award made to two European cities, was developed in 1985 and, according to the EU Commission's ECoC factsheet 'Being a European Capital of Culture brings fresh life to these cities, boosting their cultural, social and economic development' (EU 2020, 1). Held for one year, the award involves a programme of cultural events in the host city. Intrinsically bound up with this, certainly since the Glasgow 1990 ECoC, is the concept of urban regeneration and the (re)marketing of elements of a city (e.g. as more creative, more investment worthy or a tourist destination). Cities may adopt a different emphasis and not all take the same route. Fitjar,

Rommetvedt, and Berg (2013), for example, discuss how the host city Stavenger had the mission 'to focus firmly on culture which is in complete juxtaposition with its partner ECOC city, Liverpool, whose bid was based on economic regeneration' (2013: 66). Largely though, as Boland (2010) points out, ECoC is indicative of policy approaches that utilize culture for economic development and propagate a notion of regeneration via cultural programmes. In terms of the expectations of hosting ECoC, the associated narratives of regeneration or culture or creativity as a driver are ever present (see Degen and Garcia 2012, here in relation to the 'Barcelona Model'). This also has impacts on the built environment, with new structures or changes to existing sites. As Jones and Ponzini (2018) assert, mega-events can often lead to the bypassing of usual procedures in terms of changes to the built environment or the management of heritage sites (434), highlighting a gap in both literature and policy which might examine the often delicate balance between heritage preservation and mega-event planning.

Tensions between social, cultural and economic development have been explored within a wider and oft critiqued shift of the cultural policy paradigm towards the creative industries. Especially in, though not confined to, the U.K. this was framed by New Labour policy of the late 1990s onwards, and the establishment of the Department of Media, Culture and Sport and a particular vision and policy relating to a newly (re) defined cultural and creative industries and creative economy. This focus on creative industries as significant to the economy also coincided, with and encompassed shifts in, city spatial planning. The notion of creative areas in the predominantly urban setting drew upon the 'creative class' discourse (Florida 2002) and creative clustering models (e.g. from Porter 1998) which led in part to gentrification, city branding and selective regeneration policies (see Peck 2005 for an early discussion and critique). The policy and planning interventions that went along with this have and continue to be critically discussed (see Flew and Cunningham 2010; Pratt 2010; Hesmondhalgh and Pratt 2005).

There has since been a move, at least within EU Cohesion Policy and funding discourse, towards a more distinct place-based regional policy development (for a discussion of the territorial approach see McCann and Ortega-Argiles 2015; Lazzeroni et al. 2013). Here existing tangible and intangible cultural heritage assets are seen as key to a regional advantage and distinctiveness. In England, devolution towards regional governance structures and locally elected 'Metro Mayors' have shifted policy towards the more regionally relevant and distinctive, encompassing moves towards so-called inclusive growth. In the Liverpool City Region, in terms of culture, this led to a '1% Pledge for Culture' (1% of the budget), and the development of regionally distinctive initiatives (e.g. a 'Boroughs of Culture' award) that seek to redress the balance and focus away from the city centre. However, the CCIs and associated redevelopment discourse still dominates city, and increasingly town, planning and expectations, often irrespective of any critical mass of such industries.

In the context of this article, heritage is defined as something externally attributed, that is classified, listed and maintained by national or international bodies such as Historic England or UNESCO. This is not exhaustive of the meanings or values of heritage and is by no means uncomplex. What is constituted as heritage and by whom is subject to all manner of negotiations and contestations which this paper does not cover (see

Harrison 2010 for a discussion of world heritage; Smith 2006 and the discussion of heritage as externally attributed and curated; Harvey 2001 and heritage as a negotiated product of its time). As Harrison comments: 'For every object of tangible heritage there is also an intangible heritage that 'wraps' around it the language we use to describe it' (2010, 10). Within this, and especially in relation to national and world heritage, there is a tension between who defines, participates in or accesses heritage in both its tangible and intangible forms (see Beardslee 2016 and the discussion of local practices). Beyond the intangible memories and practices bound up with physical space, there are also complexities relating to those excluded from or represented within it.

Governance is here understood as a term for and means of defining, describing and analysing processes and frameworks in relation to specific subjects or policy fields, broadly defined as 'the co-ordination, regulation, or steering of affairs between actors' in a specific sector or place, such as urban governance (Obeng-Odoom 2010, 204; see also Alasuutari 2015; Hall and Taylor 1996). Within this definition, governance encompasses the relationships across sectors and networks of actors within the governance process. As Söpper (2014) stresses, there is a need to understand not only the structures of governance and their constituent parts but also the collaborative processes themselves, an identification and analysis of why actors collaborate and the rationale for networks (2014, 57). This may include the role of specific people or functions – see for example Jayne's (2012) discussion of democratically elected mayors and urban governance, or the role of individualism and leadership (see Peck 1995; also, for geographical relevance, Cocks 2013). More specifically in relation to culture, Baltà Portolés, Čopič, and Srakar (2014) explore the notion of cultural governance, suggesting that the concept 'defies precise definition' (183) and that the exploration of it requires a multidisciplinary approach, mainly due to the fact that it crosses a number of different theoretical areas. Alongside the creative industries discourse they also identify a shift in the cultural policy paradigm (and with it potential consequences). In relation to culture and cities, they highlight the increasing complexity of cultural governance, wherein cultural policy and decisions making processes are tied to the economic (see also Sacco, Tavano Blessi, and Nuccio 2009).

In order to undertake a multi-stakeholder cultural governance analysis that reflected these complexities, there was a need to explore and expand upon the perspectives of those within the governance structure itself, via qualitative network mapping in some form (see Oancea 2011; Oancea, Florez Petour, and Atkinson 2017; Wheddon and Faubert 2009). After exploration of the key sectors that had been involved in both the initial Liverpool 08 bid as well as an overview of contemporary active stakeholders within the city and city region, the cultural governance ecosystem was defined for this research as being constituent of Business and Economy, Government and Policy, University and Research, Arts and Cultural Organisation and Civil Society/Community Groups (see also literature on the quadruple helix and co-production models in terms of defining stakeholders and understanding processes and flows e.g. Colapinto and Porlezza 2012; Carayannis and Campbell 2009). These sectors and stakeholders often merged and were not exhaustive of the different actors and organizations. The majority were involved in the '08 bid or events, however, the workshops purposefully included participants who had not been involved in these in order to encompass contemporary perspectives that were less likely to defend or contest certain positions or perspectives. This was an important factor, as several participants had very strong and long-held opinions on ECoC.

In terms of stakeholders, sectors and mapping participants it is important to note that, inevitably, several shifts had taken place across the decade in terms of regional and national governance, austerity and funding cuts to city councils. The ECoC year had coincided with the culmination of vast cultural and entertainment infrastructural regeneration such as the opening of the Liverpool One shopping centre and the Liverpool Arena and Convention Centre 2008. Key economic organizations during the bidding process shifted in name or merged in function – for example the Arts, Culture and Media Enterprise, The Mersey Partnership and the North West Regional Development Agency no longer exist, merging instead into organizations such as Liverpool Vision, or replaced by new structures such as the Local Enterprise Partnership (LEP). The election of a city mayor in 2012 was followed in 2017 by that of a Metro Mayor for the devolved Liverpool City Region (LCR). Some networks that were part of '08 demonstrated a continuity, e.g. LARC (Liverpool Arts Regeneration Consortium), COol (Creative Organisations of Liverpool, developed from SMAC, Small and Medium Arts Collective), both discussed below and LCC's Culture Liverpool department. LCC had set up the Liverpool Culture Company to co-ordinate and deliver ECoC (as the managing and commissioning body). The Culture Company was replaced by Culture Liverpool in 2009.

Three 'snapshot' map visualizations were created for the workshops – 2008, 2013, 2017/18 (see Image 1). These included organizations, events, policies or networks within the local, regional, national and international scale. These were in no way complete, rather they were merely intended to initiate debate around what or who might be missing. In addition to group discussions, participants were asked to contribute post-it notes highlighting any key shifts or concerns and to draw their own network maps noting any details specific to their sector or experience that were not covered in the workshops (see image 2, example of a network map drawn by the participant; image 3, post-it notes placed on the map) (Figures 1 and 2).

A total of 8 mapping workshops were undertaken in 2018, with a total of 44 participants drawn from the identified ecosystem. Of these, three took place at the University of Liverpool. These included: Liverpool Local Enterprise Partnership; MerseyCare (National Health Services Foundation); Liverpool Bid Company, Liverpool Arena Conference Centre; Baltic Creative CIC; the Fabric Quarter development; local universities (Liverpool Hope University, John Moores University, University of Liverpool); British Music Experience; Dadafest, a disability arts organization. Two arts and culture sector-specific workshops took place within prescheduled meetings. The first was a LARC meeting where most of the city's larger arts and cultural organizations were present (the Bluecoat Gallery, Everyman and Playhouse Theatres, National Museums Liverpool (NML), Liverpool Philharmonic, Culture Liverpool, FACT, the Unity Theatre; Culture Liverpool also present). The second was at a COol meeting which included 18 participants drawn from smaller independent arts and cultural groups in the city and city region. Three individual workshops were undertaken at stakeholder sites (NML, LCC and Liverpool Vision).

An interview with Engage, a Liverpool Waterfront Residents Association and Community Interest Company (CIC) was also conducted in 2018 and is drawn

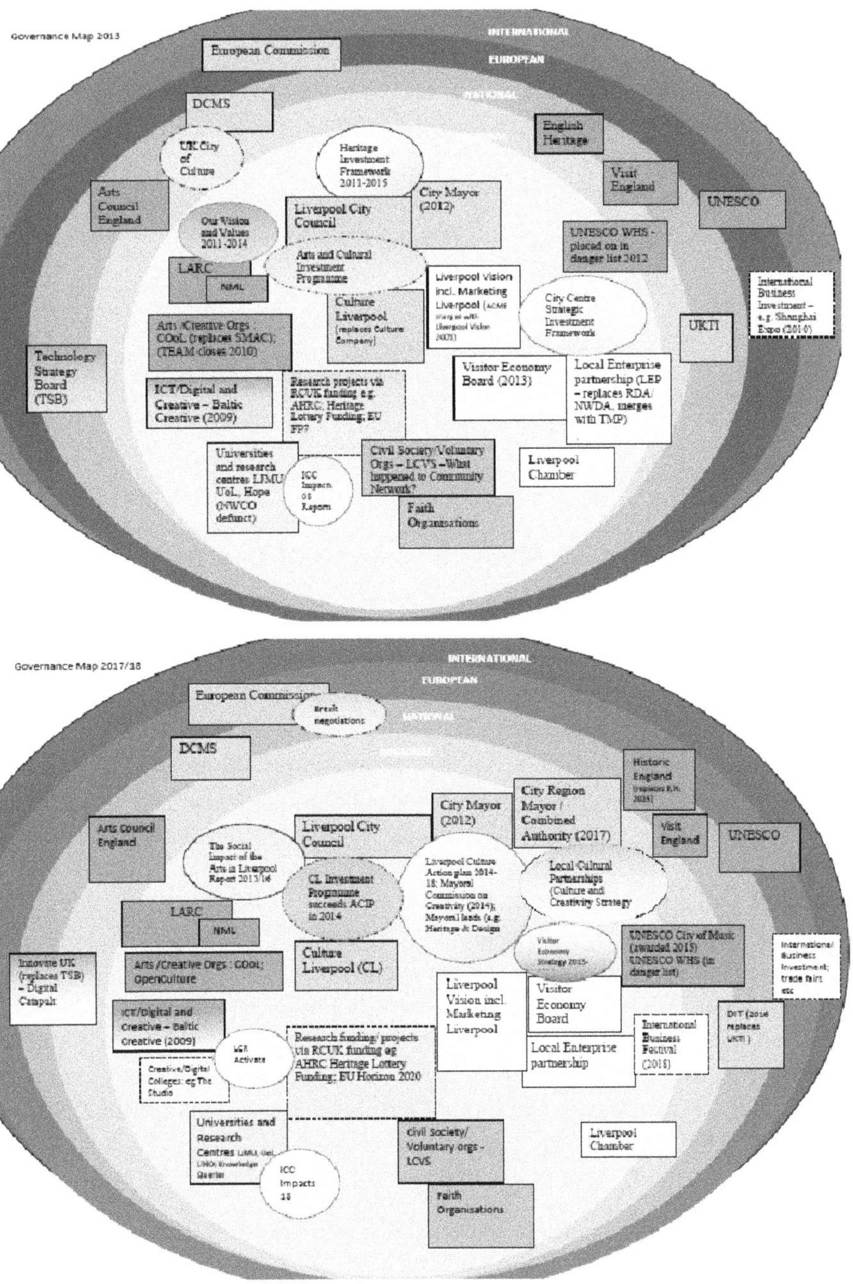

Figure 1. Example of two 'working maps' (2013, 2017).
Source: Author's own illustration.

upon. Participant observation at four city centre debates on the WHS (2017/2018), attendance at a 'City Branding' event (2017) and a large anniversary symposium (2018) are also explored here, as is a content analysis of the use of culture, and heritage, as a theme or as a thematic area within official strategy or policy documents from 2008 to 2018.

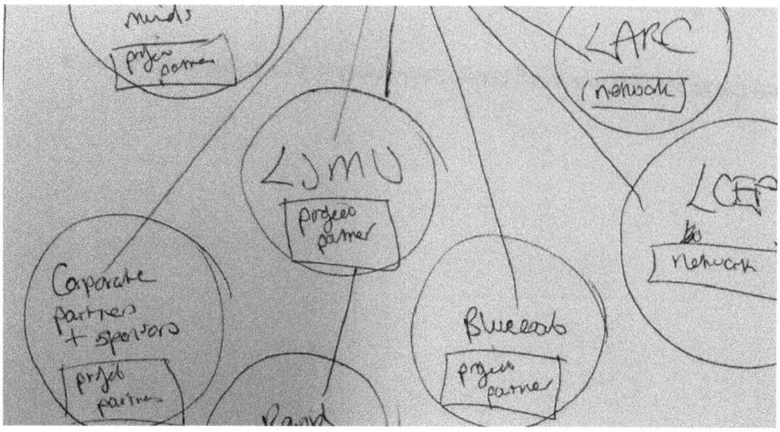

Figure 2. Participant network map, detail.

3. Overview: the ECoC legacy in the context of Liverpool

3.1 Evaluation and '08

In terms of ECoC, definitions and evaluations of impact and legacy can be problematic. Some measurable outcomes, for example, tourism, tend to peak and then even out, as Falk and Hagsten (2017) discuss in relation to overnight stays in ECoC hosting cities (see also Jones and Ponzini 2018, 437). In general, there is, as Fitjar, Rommetvedt, and Berg (2013) comment, a lack of rigorous evaluations of social outcomes across mega-events. Whilst hosting a mega-event such as ECoC may attract investments and improvements to infrastructure, it is the intangible legacy that is a main outcome – and this is, as Ferrari and Guala (2017) discuss, far more difficult to identify or measure. Much also depends, as Németh (2016) points out, on the governance networks (and power relationships) developed prior to and during the event.

For Liverpool '08 several claims were made, but little was – or can be – tangibly linked to the hosting of ECoC despite a large-scale Impacts '08 evaluation programme commissioned by the city council. Connolly (2013) points out that projected economic returns for Liverpool 08 'were rarely questioned and Impacts '08 fails to present any detailed breakdown of the numbers or types of jobs created by Liverpool's hosting of COC08' (173). He argues:

> Because the cultural planning model offered by Liverpool was theoretically incompatible it was, subsequently, practically undeliverable. Before the 'Liverpool model' can be assessed, it is essential that there is clarity as to what this 'model' actually is. Part of this is to clarify why Liverpool abandoned the approach which won the city its bid, and on which its initial planning was based. The academic study of Liverpool 08's long-term benefit – Impacts08 – fails to do this in any way. (178)

The commissioning of the research by LCC arguably led to an expectation and a usage- and reusage – of a selection of the results. For example, whilst the Liverpool 08 evaluation concentrated on issues such as representation (city branding) and perceptions, and economy, a greater understanding of the usage of existing heritage sites was not an overriding concern. The question of what evaluation can reasonably cover (and indeed evidence) is important, as are the questions of what is left out and why. This is not least because evaluations such as those used for Liverpool significantly affect how understandings and expectations of ECoC impact continue to be drawn upon, for example in current U.K. City of Culture models or by other prospective ECoC hosting cities, and because these have substantial impact on how local polices are created, justified and implemented.

Literature has in turn extolled and criticized the lead up to, hosting and evaluation of '08. Some take a critical approach in relation to the event and/or the evaluation (see Boland 2010); others utilize the event and indeed the Impacts 08 data as a gateway or overview (see Liu 2016) or expand on research undertaken during the Impacts'08 programme (see O'Brien 2010). Arguably the existence of a commissioned evaluation programme contributed to Liverpool '08 being discussed as something of its own case. Another reason for the highly case-study-centric approach is the perceived uniqueness of the case as a city that underwent many well-documented years of decline. Uniqueness and specificity are key to understanding processes, however, they should not add to an already problematic legacy narrative which, when based on exceptionalism, becomes difficult to constructively unpick. More recently and informatively, several studies have gone beyond this and drawn upon thematically relevant international case study cities (see Jones 2020, discussed below; Urbančíková 2018), thereby understanding and critiquing approaches, expectations or divergences across different examples.

3.2 Liverpool: city context

Most overviews of Liverpool begin with phrases similar to 'once a great port city'. Its waterfront location on the North West of England made it a centre for commercial trading during the British Empire and this aspect of the city's past forms the basis of its UNESCO World Heritage Status. The shipping industry shifted and declined in the twentieth century, and with it Liverpool's docks and related industries. The city was bombed during the Second World War, with large-scale damage caused to several areas. There followed a well-documented post war decline: widespread unemployment, depopulation, riots and the move towards a proposed 'managed decline' of the Thatcher era. Through interventions by the then Minister for Merseyside various public-private interventions and regeneration initiatives ensued. These initiatives were a cornerstone in subsequent understandings of entrepreneurial regeneration in the city and region, based on property-led regeneration, competition and private investment. Large public funds were also funnelled into the city via European Council funding from the 1990s onwards (EU Objective II and Objective I funding).

Sykes and colleague's (2013) city profile of Liverpool explores the city within the context of national averages and indices of multiple deprivation. These highlight continuing challenges in terms of education, health and employment. The authors comment on the narrative of regeneration which 'has become the city's dominant, if seldom quantified or questioned, objective' (2013, 300). The emphasis on regeneration still prevails, overshadowing the need to develop existing sites or initiatives. As Mulhearn and Franco (2018) in their paper exploring the boom in purpose-built student accommodation in Liverpool comment 'Objective 1 status is no longer justified given that local GDP levels have closed on the EU average. In the EU's terminology, the locality needs to now think about strategies to consolidate growth rather than engender recovery' (2018, 480).

Whilst Liverpool is no longer ranked as the most deprived local authority in England, in the 2019 indices of deprivation the city was the 4th most deprived local authority in England (by rank; 3rd by average score) and has the 2nd highest number of areas in the most deprived 10% (English Indices of Deprivation, statistical release, Sept 2019). Indices and national averages are important to situate and understand the challenges which Liverpool has faced and still faces. They are also key in assessing the veracity of the wider claims for cultural regeneration and impacts on participation, health, tourism and on wider economic activity. Here, as Boland (2010) comments 'notwithstanding a flood of investment and a period of economic growth Liverpool remains a polarised city on key socio-economic indicators (e.g. jobs, unemployment, incapacity benefits, housing and health') which 'fail to chime with the rhetoric of Liverpool as a renaissance city' (636). This echoes Connolly's (2013) comments in relation to Glasgow's ECoC hosting that 'The espoused economic successes of Glasgow stand in stark contrast to contemporary health and poverty statistics which suggest that the approach adopted by the city has resulted in social and economic polarization'. (2013, 169)

Imbalances in representation and power relationships furthermore negatively affect inclusive development strategies. In the case of Liverpool, these power relationships pre-dated ECoC and, in the words of one economic sector stakeholder consulted during the research workshops, fitted into 'a dysfunctional space'. This is of course not singular to Liverpool. As Németh (2016) points out:

> A mega-event, and especially a cultural one such as the ECOC, has the potential to reach and mobilize various segments of the society and to bring to the surface diverse constellations of both openly declared and internalized power relations. (Németh 2016, 55)

3.3 Heritage, planning and the 'Plaque on the Wall'

For Liverpool heritage necessarily includes colonial histories, the city having played a central role in British Imperialism and the slave trade- attested to in its UNESCO listing and explored in its museums, for example in NML's International Slavery Museum and the Museum of Liverpool (see Arnold de Simine 2012 and Seaton 2001 for a discussion of heritage tourism in relation to slavery).

The Liverpool Maritime Mercantile City WHS, constituent of six areas and associated buildings in the historic centre and docklands, was listed by UNESCO in 2004 under Criteria II, II and IV (see http//whc.unesco.org/en/list/1150). The Outstanding Universal Value (OUV) of the site – what defines it as worthy of world heritage status according to UNESCOs criteria – is the significance of Liverpool as a major world trading centre and port in the eighteenth, nineteenth and early twentieth centuries. At the time of the ECoC bid, Liverpool's WHS was on the 'tentative list' awaiting UNESCO designation. It was present in the ECoC bid document (Liverpool Culture Company 2004) as a case study and would go on to form a small part of the Liverpool '08 event programme (via the Creative Communities, Heritage and Welcome Team) with a number of educational and community-focused events. It was never embedded within the ECoC celebrations and did not involve local heritage actors in any significant way. As Jones (2020), using case study comparison sites of heritage and ECoC in Genoa, Istanbul and Liverpool, comments

> The city planned a massive city regeneration program around the event, but the city's heritage, of which there is a significant amount, did not take on a central role and was largely left out of the narrative of the event.' (115)

This was unlike the two other case study cities where advocates and conservation official were involved. In Liverpool no key heritage actors worked alongside the Liverpool Culture Company (178)

Often, as Jones and Ponzini (2018) comment, there can be a disconnect between mega-event planning and urban heritage, partly due to the difficulty in merging the two policies, with conflicting goals and 'certain tunnel vision in both fields' (442/3). In Liverpool's case, there was an existent complex issue of cultural heritage buildings and sites and regeneration. Indeed, it commenced soon after the Second World War. As Rodwell points out there was

> Disuse and decay in the historic docklands, untreated bomb damage, and planning blight resulting from misguided and destructive redevelopment schemes combined by the 1990s to leave a legacy of serious scars and lack of coherence in the city's urban geography. (2014: 22)

More recently, a boom in purpose-built student accommodation (PBSA) in the city, has endangered or demolished several buildings that were locally and architecturally significant. Mulhearn and Franco (2018), writing on PBSA in the city, discuss the scale of the property developments. Two high rise blocks on Lime Street, alone providing an additional 1500 beds, involved the

> fiercely contested demolition of a pre-First World War cinema front age and neighbouring buildings also of architectural interest (...) Other objections include the potential for PBSA to undermine the aesthetics of an area that still retains a number of notable nineteenth century buildings that are slowly being brought back into use. (2018, 489)

A further complexity in the redevelopment narrative that Liverpool had become wedded to is the issue of the city's WHS and with it the need to balance conservation of the urban heritage landscape with the drive to regenerate areas of the city through wider spatial planning frameworks. In 2012, UNESCO placed Liverpool on the list of 'World Heritage in Danger' as a consequence of the granting of planning permission for the proposed construction of 'Liverpool Waters', a redevelopment of the docklands including several multi-storey buildings. The planned redevelopment by the Peel Group is vast in scale and falls within the WHS and its surrounding buffer zone. Several of the originally proposed buildings were high rise and would have altered the skyline and visual landscape of the waterfront – by fragmenting the dock areas and thereby – UNESCO argued, and this would affect the Outstanding Universal Value (OUV) and authenticity of the site. Concerns were also issued regarding the lack of overall management of new developments and lack of awareness on the part of developers and the wider public about the WHS. In June 2017 UNESCO issued a final warning: if planning proposals were not addressed Liverpool would lose WHS status by 2018.

It has been reported several times that the city's major has referred WHS status as a plaque on the wall, or, as in the below newspaper reports written some four years apart, a certificate in the town hall:

> Cllr Anderson made clear his priorities for 2012. He said: "Turning Peel Holdings (the developer) away doesn't say to the world that Liverpool is a thriving modern city. It says

> we're a city that is stuck in the past (...) The Lib-Dems would turn away 20,000 jobs and £5bn of regeneration, all for the sake of a certificate on the wall in the Town Hall. I say that's madness. Whatever happens in 2012, let me be absolutely clear about one thing: we will back Liverpool Waters. (Liverpool Echo 2012)

> The mayor of Liverpool has dismissed the concerns of "luddite" campaigners who warned a new development could cost the city its UNESCO world heritage status, claiming a "certificate on the wall" did little to boost the area's prospects. (Perraudin 2016)

It should be noted though that whilst the waterfront development plans were the problematic aspect (or outcome) there were issues with the initial listing (for a discussion of this, see Rodwell 2008; Gaillard and Rodwell 2015) and with the communication of the site's importance locally. A report conducted in 2013–2014 (ICC 2014) found that the site was neither locally embedded (its WHS values were poorly communicated) nor was it valued or marketed by the city. The lack of communication is significant. As Schoch (2014), in relation to the delisting of the Dresden Elbe Valley, comments there is a need for implementation and co-ordination of the meaning and requirements of the UNESCO convention across all levels. This is something that never happened in Liverpool.

4. From 08 to 18: findings from the Qualitative data

4.1 Whose World Heritage? Discussions around delisting

In the mapping workshops, one cultural stakeholder commented that WHS designation: 'was in the long term of things possibly more important than being the ECoC but it was seen as politically dispensable until recently'. This recent shift towards some sort of action or response only occurred when the WHS faced de-listing.

Engage Liverpool, the aforementioned resident's association, who as Speake and Pentaraki (2017) contend have 'an important and to-date, under-reported, story to tell of the encouragement and facilitation of city centre residents' voices' (46), was central in highlighting the need for action. Under the title 'A Status Worth Fighting For' they staged three seminars in the city in 2017 (followed by three more in 2018). Key UNESCO representatives were invited (UNESCO's voice had until then not been heard in the city), alongside local policy makers, academics and property developers. Also present were representatives or examples from other WHS heritage sites. The seminars, each held in buildings within the WHS, highlighted the fact that until this point little to no open discussion had taken place. Questions from the public were invited and in the first seminar (Oct 2017) the first question posed to the panel asked why the council allowed development to go ahead even after UNESCO's 2012 warning (to which the councillor responded that for a city that had witnessed several years of economic decline the old buildings represented a success and that there was a need to try and build that again in a new way). Another local resident commented that she didn't understand why 'generic, glass fronted' buildings were being built, instead she wanted 'people to come to Liverpool and know they are in Liverpool'. Throughout the event attendees were given time to talk to each and share their thoughts– these ranged from concerns about perception of the city through to the impact of the waterfront development on the wider environment. There was, however, little to no coverage of the seminars in the local press.

The LCC did respond to the threat of de-listing via the appointment of a 'World Heritage Taskforce', a 'Mayoral Lead for Heritage' and the submission of a Desired State of Conservation Report (DSOCR) to UNESCO. Based on the DSOCR, the 42nd Session of the UNESCO World Heritage Committee decided the site would not be delisted, it would however remain on the World Heritage in Danger list.[3] The DSOCR had promised a new interpretation and communication strategy, measures to reinforce planning permission procedure, a 'skyline' policy and, crucially, the increasing engagement of civil society, via Engage Liverpool CIC and Merseyside Civic Society. It should be noted that in the UNESCO draft decision it was stated that whilst there was a positive move to safeguard the site, several measures in the DSOCR were not complete or relied on documents still in preparation (see UNESCO 2018, 4–7). Whilst these did suggest some shifts, they do not appear to have fundamentally altered the city council's actions or perspectives. Plans remain in place for the building of a stadium on the waterfront which would still present issues in terms of the long-term fate of the WHS.

4.2 A world city? Enduring narratives of competition

In LCC and LCR documents, statements and presentations after ECoC (2008–2018), there remained a strong emphasis on how the city must become a more outward-looking international place rather than concentrating on local issues. Here world heritage, and other designated heritage buildings, are framed as a local or internal issue, the concerns of 'luddite' campaigners (as in the above-cited newspaper). Heritage becomes valuable only when situated within the economy, else it remains seen as something that won't yield. A world city is in effect here purely understood as one which invites international investment and development.

Across several documents from 2008 to 2018, there is a consistent use of the notion of a global or world city, at times as a preferred or binary option i.e. the city sees opportunities and its role as more international than national. Keywords replicate those made in strategies and publications from bid to hosting year and include repetitions of phrases such as 'global city' 'international' 'world city' 'ambitious in focus' and 'Liverpool as destination'.[4] This was further explored in a city branding exercise that took place in 2017–2018 which suggested that 'The narrative should also recognise and reflect the extraordinary pace and scale of the changes taking place in the global economy - and show that Liverpool is at the forefront of many of them'. (Heseltine Institute, Liverpool Branding Report to Marketing Liverpool 2016, 6)

Comments made by LCC and Culture Liverpool representatives in focus groups undertaken in 2016 also underlined the opinion that 'Liverpool is its own thing' and that the city is internationally known and there is no need for comparison with other U.K. cities.[5] Similarly, a key LCR representative interviewed in 2018 in relation to the Liverpool 08 anniversary commented: 'we're a global brand; an international player' and that there is a need to 'set ourselves against what's happening in Rome, Barcelona – not down the road' (Workshop, 2018). This is at odds with a more contemporary regional development narrative that understands inclusive development beyond the city centre and periphery. Distinctiveness is conflated with marketability rather than embedded internal (local) value and assets. Here the city's approach is in some ways frozen in time and remains bound to the ECoC bid (and the decades before) where city competition and redevelopment were seen as the only way forward, unheeding of Pratt's (2011) warning to cities: 'to escape from the limitations of 'cookbook' approaches to creative cities based upon narrow branding or place marketing (…) consider instead a more nuanced approach that is sensitive to local cultures and differences' (Pratt 2011, 123).

4.3 The ECoC legacy narrative

> Culture is the city's USP [Unique Selling Proposition] and has made Liverpool globally famous (…) It is the rocket fuel for regeneration. (LCC, Inclusive Growth Plan 2018, 81)
> We[6] made culture the driver for regeneration and development so we instantly connected culture to the market economy (…) we saw culture through the prism of 'how can we get more investment?', with the discussion about sustainability or what was best never really taking place. (Community Stakeholder, Workshop 2018)
> The balance between the social and economic since 08, a lot of these organisations are economic outward facing organizations yet actually ECoC was about creating more opportunities for the 3rd party social sector that I think was probably closer in 08 and now is wider. If '08 was all about these organisations (…) you would hope that by 2013 those parts of the city that were more disaffected before 08 would have become more integrated, empowered within that space but they've probably become less empowered, less relevant in that space. (Economic Organisation Stakeholder, Workshop 2018)

In recent LCC and LCR publications and planning documents, there is an emphasis on the importance of culture. However, this understanding of culture is as a socio-economic driver, and often equated with the ECoC legacy of large events or touristic offer. Here '08 is cited as a catalyst and as 'a triumphant year' (LCC Inclusive Growth Plan 2018), and in many ways the understanding and definition of culture never deviates much from this ECoC framing. It is continuously linked to the wider narrative of regeneration with statements such as 'culture as a catalyst for urban renaissance' (83) and as 'the driver for spectacular growth within the visitor and night time economy' (81). As such, there remains a consistent alignment of culture with economic growth, demonstrated by the often-repeated slogan of the city's mayor that culture is the 'rocket fuel for regeneration'. However, as one business sector stakeholder demonstrates, this is met with some criticism:

> It's that view, isn't it, that culture is the rocket fuel of regeneration, so we keep getting told, which is great and we have big events but actually the grass roots in the city and the cultural organisations where's the leadership for them and where's the support for them?. (Mapping Workshop 2018)

Grand claims for the economic values of culture now extend to heritage in both its tangible and intangible form. Whilst the wider WHS is rarely mentioned in LCC or LCR documents, the famous docks and the intangible music and sporting heritage of the city are viewed as the catalyst for inward investment, large-scale redevelopment and economic development. Here heritage is an extension of culture as asset – it is externally marketable and must be harnessed:

> To stay ahead, Liverpool needs to continually renew its offer and sustain investment in its heritage and visitor assets, deliver transformative infrastructure projects and innovative cultural programming. (LCC, Inclusive Growth Plan 2018, 82)

Discussions with local government actors during the workshops highlighted the fact that 'the cultural' is seen as informing strategies beyond arts and culture and is perceived to be a marketable asset that leverages external investment and tourism. Here the promise is of 'economic growth through culture and creativity', which are seen as central drivers for the regional economy (LCR Culture and Creativity Strategy and Framework Draft for Consultation 2018, 15).

Related to this, a contrasting issue across all workshops was the role of large events within the city's cultural offer and strategy development and a perceived lack of support for grass roots organizations. For several community artists and cultural groups, there was a shared concern that funds might have instead been steered towards more local or potentially more lasting initiatives. This was at the time also framed by the closure of Hope St, a community arts venue that had formed the original meeting place for COoL (at that point SMAC) during 2008. One example of this can be seen in the individual post note exercise undertaken in the COol workshop. Here a local arts stakeholder wrote: 'Commercial big scale events – 'smaller' events not then as important if not VISIBLE'; 'Local Authority Arts Officers being cut; little arts commissions from the city/less open calls' (Image 4) (Figure 3).

For LCC and Culture Liverpool representatives, the focus on large events is presented as positive. It is seen as proof of the city's confidence in bidding for and hosting large-scale events or awards and building on ECoC experience and networks. The words 'greater confidence' are peppered across interviews with LCC stakeholders, drawing upon their expertise in the production and delivery of large events. The city's top-down event-led focus is perceived by other stakeholders as more problematic. In a diplomatic exploration of this, National Museums Liverpool (NML) pointed out that in the Inclusive Growth Plan there was no real mention of the museums (other than a blockbuster Terracotta Warriors exhibition held in 2018). They suggested that in addition to one off or annual events the city should highlight the constant presence of the museums that is there for visitors as well as local communities. Others were less diplomatic. Referring to the city's approach as 'events as strategy', several stakeholders – from across cultural and

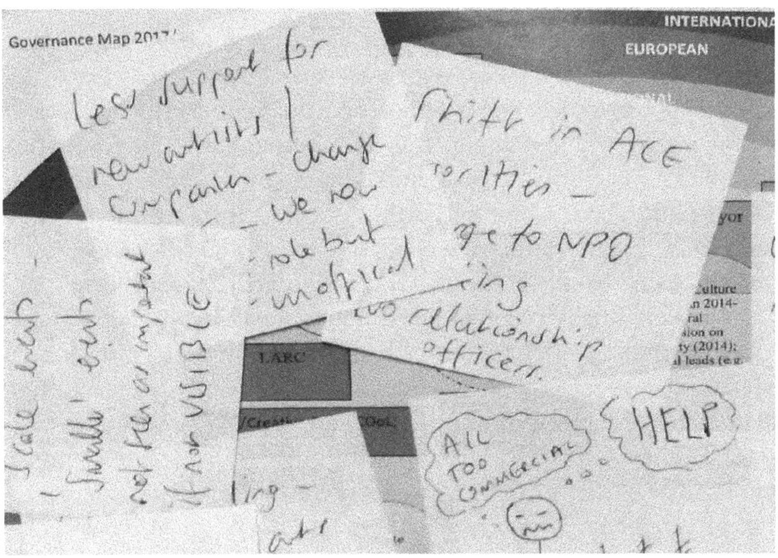

Figure 3. COoL workshop, participant responses.

economic organizations – criticized the emphasis on sourcing external companies to provide large scale events. Referring to the 'Giants' spectacle (a popular large-scale marionette parade that was held in the city in 2012, 2014 and then in the anniversary year 2018) one stakeholder commented:

> You can't just keep getting the f***ing Giants back year after year and saying that's what we do, that's what Liverpool's famous for. How long will that last for until the penny drops that anybody that has money can buy that stuff in. (Mapping workshop, 2018)

5. Conclusion: 'It's very interesting to me, what have we really become since 08?'

This question, posed during one of the research workshops, moves the article towards a reflective conclusion that returns to the issue of legacy narratives. In 2006, an audit commission report into LCC Cultural Services referred to an uncertain future, concluding that cultural planning was dominated by ECoC. It warned that it was not clear what the city council and their partners meant by ambitions to be 'world class' or a 'premier EU city' and found that 'within the cultural programme there is a tension between the large scale 'wow' events and the need to tackle the multiple causes of deprivation at local level' (Audit Commission Report 2006: 36). Post ECoC these concerns not only remain, they have become further embedded.

The ten-year anniversary of Liverpool '08 might have presented an opportunity for a critical reappraisal, coinciding as it did with the proposed delisting of the UNESCO site, concerns about increased student accommodation building and associated issues confronting historic buildings, the loss of several community arts sites and the continuing challenges of deprivation framed by extensive funding cuts. Instead, older and to a large extent unsubstantiated claims were showcased. A lack of tangible evidence was deftly overlooked in the celebrations and symposium referred to at the beginning of this article, with statements such 'swagger', 'renaissance' or 'city confidence' reinforcing how the preferred city narrative having long since prevailed in the public discourse at the expense of other evaluations. Two years later, there is seemingly nothing that this 'swagger' and the 'rocket fuel' cannot be said to remedy.[7] A focus on visitor economy or 'wow' events that fail to take into account the different meanings and values of the city's embedded tangible and intangible heritage was further illustrated in July 2020 when the council approved plans for a city centre zipwire ride that resulted in opposition from several figures from cultural, heritage and community groups on the grounds that it would 'mutilate' the site and would cross several memorial sites and reflective spaces, in effect constituting a move towards 'Disneyfication'.[8] There remains a focus on development *despite* rather than *with* the urban heritage site. This is not singular to Liverpool, of course. As Jones and Ponzini (2018) suggest, there is too great a distance in both research and policy between mega-event studies and heritage, despite an increased overlap across the two and urgency for discussion and action. The use of mega-events as an economic and touristic intervention is also not singular to Liverpool. However, the continuing use of an ECoC success narrative – and with it an explicit reliance on and expectation of culture led (re)development – is embedded within the city's approach.

At its heart the above question 'what have we really become since '08' must remain a learning process for all cities hosting large scale events such as ECoC. A focus on the often already embedded expectations and resultant limitations of measuring impact is necessary, as is the issue of comparison. A recognition that being site-specific remains important, but that this must go hand in hand with the need for a dialogue that goes beyond notions of competition towards one that preferences shared learning rather than the promise of replicating of a perceived success. How other ECoC hosting cities utilized their heritage or foregrounded more democratic local networks is essential here not to provide templates, but to open up discussions and challenge or amplify narratives post event (see comparisons by Jones 2020; Fitjar, Rommetvedt, and Berg 2013, both discussed above, see also Öztürk and Terhorst 2012 in relation to tourism). The espoused success of Liverpool '08 and resultant propagation of a top down ECoC legacy narrative damaged a more multifaceted engagement with built heritage and with the communities and practices associated with it. This was consolidated by an initial lack of wider contextualization of the event or the selection of given effects rather than others. A blinkered emphasis was placed on an ongoing city renaissance, one which is always in progress and as such never fully assessed. From large-scale tourism-focused cultural events through to controversial property-led investment strategies, the ECoC legacy narrative was used to capitalize on the buzzwords of heritage and culture whilst simultaneously attempting to reduce them to a binary choice between progress and preservation.

Notes

1. Interview with a Liverpool Community Residents Association stakeholder, 2018
2. See the following for the use of these phrases, written at the time of the 2018 event https://www.cultureliverpool.co.uk/liverpool2018/culture-the-catalyst-of-liverpools-re-introduction-to-the-world/ and https://www.cultureliverpool.co.uk/news/liverpool-2018-ten-years-in-the-making/
3. see http://whc.unesco.org/en/sessions/42com/ for details of this meeting. Since writing this article the UNESCO World Heritage Committee met in July 2021 and made the decision to delete the Liverpool WHS- see https://whc.unesco.org/en/news/2314/. This development occurred during final corrections of the article and could no longer be included in the discussions.
4. Documents consulted include: LCC (2008) Cultural Legacy Scrutiny Panel; Liverpool First/LCC (2010) Liverpool Cultural Strategy Delivery Plan: 2010–2014 Stakeholder Summary; LCC (2011) Vision aims, priorities and values: Our Vision; LCC (2012) Report to Select Committee (Culture and Tourism) Cultural Strategy 2012/13 (LCC, committee meeting 19/06/ 2012); LCR/ LEP (November 2014) Visitor Economy Strategy and Destination Management Plan; Culture Liverpool/LCC (Undated) Liverpool Culture Action Plan 2014–2018; LCC (October 2017) Ten Streets Supplementary Planning Document; LCR (2018) Culture and Creativity Strategy Framework (Draft); LCC (March 2018) Mayor of Liverpool Inclusive Growth Plan
5. Focus group /Interview not conducted or planned by the author. Full recordings and transcription were accessed as part of this research.
6. Here the stakeholder is referring to the city as 'we'
7. See comments made, at the height of the Covid pandemic, by Culture Liverpool https://www.themj.co.uk/Liverpool-has-come-out-fighting-on-culture/218279
8. See, for example, news coverage of this https://www.bbc.co.uk/news/uk-england-merseyside-53516228

Acknowledgements

The author wishes to thank colleagues at the Institute of Cultural Capital (ICC) where this research was undertaken and funded. It formed a thematic part of a larger study, with other team members exploring different themes in relation to the decade since Liverpool 08. No final reports have yet been published at the time of writing. This paper only discusses findings from the Cultural Governance and Heritage and Planning thematic projects which were undertaken and analysed by the author, it does not draw upon any other results.

Disclosure statement

No potential conflict of interest was reported by the author(s).

References

Alasuutari, P. 2015. "The Discursive Side of New Institutionalism." *Cultural Sociology* 9 (2): 162–184. doi:10.1177/1749975514561805

Arnold de Simine, S. 2012. "The 'Moving' Image Empathy and Projection in the International Slavery Museum, Liverpool." *Journal of Educational Media, Memory, and Society* 4 (2): 23–40. doi:10.3167/jemms.2012.040203

Audit Commission Report. 2006. *Local Government Service Inspection Report.* Cultural Services LCC.

Baltà Portolés, J., V. Čopič, and A. Srakar. 2014. "Literature Review on Cultural Governance and Cities." *Kult-ur* 1 (1): 183–200. doi:10.6035/Kult-ur.2014.1.1.9

Beardslee, T. 2016. "Whom Does Heritage Empower, and Whom Does it Silence? Intangible Cultural Heritage at the Jemaa el Fnaa, Marrakech." *International Journal of Heritage Studies* 22 (2): 89–101. doi:10.1080/13527258.2015.1037333

Boland, P. 2010. "'Capital of Culture—you Must be Having a Laugh!' Challenging the Official Rhetoric of Liverpool as the 2008 European Cultural Capital." *Social & Cultural Geography* 11 (7): 627–645. doi:10.1080/14649365.2010.508562

Carayannis, E., and D. Campbell. 2009. "'Mode 3' and 'Quadruple Helix': Toward a 21st Century Fractal Innovation Ecosystem." *International Journal of Technology Management* 46 (3–4): 201–234. doi:10.1504/IJTM.2009.023374

Cocks, M. 2013. "Conceptualizing the Role of Key Individuals in Urban Governance: Cases from the Economic Regeneration of Liverpool, UK." *European Planning Studies* 21 (4): 575–595. doi:10.1080/09654313.2012.722955

Colapinto, C., and C. Porlezza. 2012. "Innovation in Creative Industries: From the Quadruple Helix Model to the Systems Theory." *Journal of the Knowledge Economy* 3: 343–353. doi:10.1007/s13132-011-0051-x

Connolly, M. 2013. "The 'Liverpool Model(s)': Cultural Planning, Liverpool and Capital of Culture 2008." *International Journal of Cultural Policy* 19 (2): 162–181. doi:10.1080/10286632.2011.638982

Degen, M., and M. Garcia. 2012. "The Transformation of the 'Barcelona Model': An Analysis of Culture, Urban Regeneration and Governance." *International Journal of Urban and Regional Research* 36 (5): 1022–1038. doi:10.1111/j.1468-2427.2012.01152.x

EU Commission. 2020. "European Capitals of Culture (Factsheet)." https://ec.europa.eu/programmes/creative-europe/sites/default/files/2020_cult ecoc_factsheet_12.08.20_en.pdf

Falk, N., and E. Hagsten. 2017. "Measuring the Impact of the European Capital of Culture Programme on Overnight Stays: Evidence for the Last two Decades." *European Planning Studies* 25 (12): 2175–2191. doi:10.1080/09654313.2017.1349738

Ferrari, S., and C. Guala. 2017. "Mega-Events and Their Legacy: Image and Tourism in Genoa, Turin and Milan." *Leisure Studies* 36 (1): 119–137. doi:10.1080/02614367.2015.1037788

Fitjar, R., H. Rommetvedt, and C. Berg. 2013. "European Capitals of Culture: Elitism or Inclusion? The Case of Stavanger2008." *International Journal of Cultural Policy* 19 (1): 63–83. doi:10.1080/10286632.2011.600755

Flew, T., and S. Cunningham. 2010. "Creative Industries After the First Decade of Debate." *The Information Society* 26 (2): 113–123. doi:10.1080/01972240903562753

Florida, R. 2002. *The Rise of the Creative Class: and how It's Transforming Work, Leisure, Community, and Everyday Life*. New York: Basic Books.

Gaillard, B., and D. Rodwell. 2015. "A Failure of Process? Comprehending the Issues Fostering Heritage Conflict in Dresden Elbe Valley and Liverpool Maritime Mercantile City World Heritage Sites." *The Historic Environment: Policy & Practice* 6 (1): 16–40. doi:10.1179/1756750515Z.00000000066

Hall, P., and R. Taylor. 1996. "Political Science and the Three new Institutionalisms." *Political Studies* 44 (5): 936–957. doi:10.1111/j.1467-9248.1996.tb00343.x

Harrison, R. 2010. "What is Heritage?" In *Understanding the Politics of Heritage*, edited by R. Harrison, 5–42. Manchester: Manchester University Press in association with the Open University.

Harvey, D. 2001. "Heritage Pasts and Heritage Presents: Temporality, Meaning and the Scope of Heritage Studies." *International Journal of Heritage Studies* 7 (4): 319–338. doi:10.1080/13581650120105534

Heseltine Institute. 2016. *The Liverpool Brand, What is it and why does it Matter?* https://www.liverpool.ac.uk/media/livacuk/publicpolicyamppractice/Liverpool,Brand,Report.pdf

Hesmondhalgh, D., and A. Pratt. 2005. "Cultural Industries and Cultural Policy." *International Journal of Cultural Policy* 11: 1–14. doi:10.1080/10286630500067598

Institute of Cultural Capital. 2014. *Heritage, Pride and Place. Exploring the contribution of World Heritage Site status to Liverpool's sense of place and future development.* http://iccliverpool.ac.uk/wp-content/uploads/2015/01/ICC-WHS-FINALREPORT.pdf

Jayne, M. 2012. "Mayors and Urban Governance: Discursive Power, Identity and Local Politics." *Social & Cultural Geography* 13 (1): 29–47. doi:10.1080/14649365.2011.635800

Jones, Z. 2020. *Cultural Mega-Events: Opportunities and Risks for Heritage Cities*. London: Routledge.

Jones, Z., and D. Ponzini. 2018. "Mega-events and the Preservation of Urban Heritage: Literature Gaps, Potential Overlaps, and a Call for Further Research." *Journal of Planning Literature* 33 (4): 433–450. doi:10.1177/0885412218779603

Lazzeroni, M., N. Bellini, G. Cortesi, and A. Loffredo. 2013. "The Territorial Approach to Cultural Economy: New Opportunities for the Development of Small Towns." *European Planning Studies* 21 (4): 452–472. doi:10.1080/09654313.2012.722920

LCR. 2018. *Culture and Creativity Strategy and Framework Draft for Consultation*. https://www.liverpoolcityregion-ca.gov.uk/wp-content/uploads/DRAFT_LCR_Culture_Creativity_Strategy.pdf

Liu, Y. 2016. "Event and Quality of Life: A Case Study of Liverpool as the 2008 European Capital of Culture." *Applied Research Quality Life* 11: 707–721. doi:10.1007/s11482-015-9391-1

Liverpool City Council. March 2018. *Mayor of Liverpool Inclusive Growth Plan- A Strong and Growing City Built on Fairness*. https://liverpool.gov.uk/council/strategies-plans-and-policies/inclusive-growth-plan/

Liverpool Culture Company. 2004. *The World in One City* (ECoC Bid Document).

Liverpool Echo. 2012. *Liverpool Council Leader Joe Anderson says City Would Sacrifice World Heritage Status for Liverpool Waters Scheme in New Year report*, Jan 2, 2012. https://www.liverpoolecho.co.uk/news/liverpool-news/liverpool-council-leader-joe-anderson-3354111

McCann, P., and R. Ortega-Argiles. 2015. "Smart Specialization, Regional Growth and Applications to European Union Cohesion Policy." *Regional Studies* 49 (8): 1291–1302. doi:10.1080/00343404.2013.799769

Ministry of Housing, Communities and Local Government. September 2019. *The English Indices of Deprivation 2019 - Statistical Release*.

Mulhearn, C., and M. Franco. 2018. "If you Build it Will They Come? The Boom in Purpose-Built Student Accommodation in Central Liverpool: Destudentification, Studentification and the Future of the City." *Local Economy* 33 (5): 477–495. doi:10.1177/0269094218792740

Németh, A. 2016. "European Capitals of Culture – Digging Deeper Into the Governance of the Mega-Event." *Territory, Politics, Governance* 4 (1): 52–74. doi:10.1080/21622671.2014.992804

Oancea, A. 2011. *Interpretations and Practices of Research Impact Across the Range of Disciplines Report*. Oxford: Oxford University.

Oancea, A., T. Florez Petour, and J. Atkinson. 2017. "Qualitative Network Analysis Tools for the Configurative Articulation of Cultural Value and Impact from Research." *Research Evaluation* 26 (4): 302–315. doi:10.1093/reseval/rvx014

Obeng-Odoom, F. 2010. "On the Origin, Meaning, and Evaluation of Urban Governance." *Norsk Geografisk Tidsskrift Norwegian Journal of Geography* 66 (4): 204–212. doi:10.1080/00291951.2012.707989

O'Brien, D. 2010. "'No Cultural Policy to Speak of' – Liverpool 2008." *Journal of Policy Research in Tourism, Leisure and Events* 2 (2): 113–128. doi:10.1080/19407963.2010.482271

Öztürk, H., and P. Terhorst. 2012. "Variety of Urban Tourism Development Trajectories: Antalya, Amsterdam and Liverpool Compared." *European Planning Studies* 20 (4): 665–683. doi:10.1080/09654313.2012.665037

Peck, J. 1995. "Moving and shaking: business élites, state localism and urban privatism." *Progress in Human Geography* 19 (1): 16–46. doi:10.1177/030913259501900102

Peck, J. 2005. "Struggling with the Creative Class." *International Journal of Urban and Regional Research* 2 (4): 740–770. doi:10.1111/j.1468-2427.2005.00620.x

Perraudin, F. 2016. "Opponents of New Liverpool Tower Block are Luddites, says Mayor." *The Guardian*, October 12. https://www.theguardian.com/uk-news/2016/oct/12/opponents-of-new-liverpool-tower-block-are-luddites-says-mayor

Porter, M. 1998. "Clusters and the New Economics of Competition." *Harvard Business Review* 76 (6): 77–90. https://hbr.org/1998/11/clusters-and-the-new-economics-of-competition

Pratt, A. 2010. "Creative Cities: Tensions Within and Between Social, Cultural and Economic Development. A Critical Reading of the UK Experience." *City, Culture and Society* 1: 13–20. doi:10.1016/j.ccs.2010.04.001

Pratt, A. 2011. "The Cultural Contradictions of the Creative City." *City, Culture and Society* 2 (3): 123–130. doi:10.1016/j.ccs.2011.08.002

Rodwell, D. 2014. "Negative impacts of World Heritage Branding: Liverpool - an unfolding tragedy?" In *Between dream and reality: Debating the impact of World Heritage Listing. Primitive Tider Special Edition.* edited by Hølleland, Herdis and Solheim, Steinar, 19–34. Oslo: Reprosentralen.

Rodwell, D. 2008. "Urban Regeneration and the Management of Change." *Journal of Architectural Conservation* 14 (2): 83–106. doi:10.1080/13556207.2008.10785025

Sacco, P., G. Tavano Blessi, and M. Nuccio. 2009. "Cultural Policies and Local Planning Strategies: What Is the Role of Culture in Local Sustainable Development?" *The Journal of Arts Management, Law, and Society* 39 (1): 45–64. doi:10.3200/JAML.39.1.45-64

Schoch, D. 2014. "Whose World Heritage? Dresden's Waldschlößchen Bridge and UNESCO's Delisting of the Dresden Elbe Valley." *International Journal of Cultural Property* 2: 199–223. doi:10.1017/S094073911400006X

Seaton, A. 2001. "Sources of Slavery-Destinations of Slavery." *International Journal of Hospitality & Tourism Administration* 2 (3-4): 107–129. doi:10.1300/J149v02n03_05

Smith, L. 2006. *Uses of Heritage*. New York: Routledge.

Söpper, K. 2014. "Governance and Culture – a New Approach to Understanding Structures of Collaboration." *European Spatial Research and Policy* 21 (1): 53–64. doi:10.2478/esrp-2014-0005

Speake, J., and M. Pentaraki. 2017. "Living (in) the City Centre, Neoliberal Urbanism, Engage Liverpool and Citizen Engagement with Urban Change in Liverpool, UK." *Human Geographies* 11 (1): 41–63. http://dx.doi.org/10.5719/hgeo.2017.111.3

Sykes, O., J. Brown, M. Cocks, D. Shaw, and C. Couch. 2013. "A City Profile of Liverpool." *Cities* 35: 299–318. doi:10.1016/j.cities.2013.03.013

UNESCO. 2018. *World Heritage 42 COM* (WHC/18/42.COM/7A.Add), Paris 28th May 2018. http://whc.unesco.org/archive/2018/whc18-42com-7AAdd-en.pdf

Urbančíková, N. November 2018. "European Capitals of Culture: What are Their Individualities?" *Theoretical & Empirical Researches in Urban Management* 13 (4): 43–55.

Wheddon, J., and J. Faubert. 2009. "Framing Experience: Concept Maps, Mind Maps, and Data Collection in Qualitative Research." *International Journal of Qualitative Methods* 8 (3): 68–83. doi:10.1177/160940690900800307

Internationalizing small-sized cities through mega-events: the case of Matera-Basilicata 2019 European Capital of Culture

Marina Rotolo

ABSTRACT
The paper is part of an ongoing thesis on the city of Matera (Italy). The research questions the effects and challenges of a double international recognition on urban space production: the mega-event 'European Capital of Culture' and the UNESCO World Heritage status. Using Matera's bid book to become the European Capital of Culture, the paper focuses on the vocabulary used by the experts in charge of the application and questions the city models underlying the cultural programme. Through the use of an entrepreneurial vocabulary, the stakeholders of the Matera–Basilicata 2019 Foundation propose to transform the classic modes of bureaucratic management and the way of conceiving urban design. In this perspective, the challenge is no longer to invest in infrastructural components or public facilities but to create a new attractive image of the city. In this context, the Foundation in charge of the organization of the event acts as a company to promote new narratives and to provide new services for the city. The label 'European Capital of Culture' is therefore used to promote the city through a new urban vision, which reinterprets the local and historical urban representations.

The label as a policy tool

European Capital of Culture,[1] UNESCO World Heritage List, Olympic Games … Over the last few years, a wide range of designations has been created in the cultural, heritage and tourist world. For political science researchers, this increase of labels has been interpreted as a sign of the transformation of public action, in order to develop the attractiveness of cities and, through this process, gain not only visibility but also attract private investors (Béal, Epstein, and Pinson 2015). The aesthetic qualities of territory are no longer rewarded, but the way in which political and economic stakeholders manage it. According to the sociologist Renaud Epstein, these meritocratic instruments (labels, prizes and brands) have become tools of public administration, useful for fostering urban innovation and its dissemination in the national or even international sphere (Epstein 2013). Through study trips between professionals, congresses and trade fairs, networks of cities are being set up. These networks allow the circulation of 'good practices', which are considered exemplary. In this dynamic, the State presents itself as a guarantor of the regulation and control of urban policies through the label. The title thus legitimizes the competition between territories and acts as a driving force for the development of cities and their future projection. Innovative local experiences can thus be identified, validated and promoted at the national level. Participation in this competition leads to the emergence of urban models policies that can be duplicated in other contexts and circulate across labels (Bourdin and Idt 2016; Arab 2007). In this perspective, Heritage and Culture are often being used to produce new urban images (Vanolo 2008; AlSayyad 2013) and develop regeneration projects (Bianchini and Parkinson 1994; Evans 2001; Zukin 2009). These strategies question the idea of 'homogenization' not only of urban landscapes (Zukin 2012) but also the relation between heritage and cultural policy (Jones 2020; Jones and Ponzini 2018). In this context, cities are becoming 'products' to be marketed in order to attract tourists, residents, commercial activities and investments (Ashworth and Voogd 1990; Russo 2002). These cultural policies models also tend to attract a new kind of professionals, theorized by Richard Florida as a 'creative class'[2] (Florida 2002; Darchen and Tremblay 2008), which concept is discussed by several authors (Bourdin 2005; Ponzini and Rossi 2010). The main challenge for local authorities would therefore lie in their ability to attract these individuals through a cultural offer and a 'bohemian atmosphere' of the territory (Ambrosino 2019). Participating in the creation of a new perception of the city, the production of tourist images within the framework of the European Capitals of Culture (Richards and Wilson 2004), is coupled with an urban renewal strategy often following the principles of the creative city (Bianchini and Parkinson 1995).

In the case of this European cultural label, this shift from marketing methods to a tool for public and political action is clearly observed. Two examples of cities are indicative of this transformation: Glasgow in 1990 and Liverpool in 2008. The Scottish city introduced a major turning point in 1990 in the 'European Cities of Culture' programme, since it was the subject of a personal decision by Margaret Thatcher[3] to transform the old industrial city. As Renaud Denuit explains in his inventory of ECoCs, this choice aroused indignation within the European Commission at the time: 'What was "cultural" about Glasgow compared to the beautiful cities[4] that had received the title until then? The Iron Lady had made a domestic policy decision, but left her European partners no choice but to endorse it' (Denuit 2018, 23). The political strategy of the United Kingdom then re-questioned the programme, financed by the budget of the European Community but depending on operations decided according to national criteria for economic purposes. Nearly two decades later, Prime Minister Tony Blair repeated the operation by favouring Liverpool over other candidate cities for the ECoC 2008. According to Renaud Denuit, the internal competition was distorted with exclusively national evaluators who do not rely on the advice of European experts. Obtaining the 'prestigious label' (Denuit 2018, 24) aimed to attract cultural and creative industries in order to revitalize the city in economic decline, but forgetting the European dimension, which is at the heart of the creation of the programme. This distancing of the State via a European label, therefore, gives it the capacity to orient the policies carried out locally without directly interfering in the decision-making of urban actors.

In order to analyse the effects of the labelling process on the urban factory, a first etymological definition of the term 'label' is necessary to question its appearance in the field of urban planning. It is in this nascent field of research that the research is included in order to study the transformation of the city under the prism of labelling. Present since 1980, in the field of marketing, the question of labelling has gradually invested studies in political science, then later in geography. This transition from marketing to the urban world reflects the use of the label as a tool for public and political action marked by the neo-liberal turn at the end of the twentieth century. Labels have thus become political instruments for enhancing the value of cities and strategic vectors for territorial transformations. In this context, the main question guides the study: to what extent does the ECoC label transform ways of thinking and projecting the city? To answer this question, the case study concerns the city of Matera, the European Capital of Culture, in 2019. Although the ECOC label concerns the whole region where the city is located ('Basilicata'), the study is focussed on the urban scale to observe the socio-spatial transformations during the preparation of the event. Sources used in this paper are based on Matera 2019s bid book,[5] but also data collected during different field surveys.[6] Both sources are mobilized in order to understand what kind of urban visions this label brings and to what extent these visions correspond to a phenomenon of internationalization of the urban factory. More generally, the aim is to analyse the dialectic between a global scale and a local context in southern Italy. To achieve this, an inventory of anglicisms present in the Italian bid book was created to focus the attention on language practice, and more specifically, on the experts' vocabulary (Coudroy de Lille and Ratouis 2019; Fijalkow 2017). Indicating the number of times these English expressions appeared in the text, a lexicon was created to explain their context of enunciation, their formatting (italics, between quotation marks, bold), their meaning and finally, their classification by theme. Using this method, the aim is to show that this specific language reflects a new way of conceiving urban space production and its transformation through international concepts. Conveying a more global way of thinking about territories, this specialized language translates alternative management methods that favour specific skills and social classes while disseminating a doctrine on cultural and urban policies to be implemented.

Questioning the notion of 'label'

But what exactly is meant by the term 'label'? The dictionary of the 'Académie Française' tells us about the origin of this word borrowed from the English word 'label', translated as 'étiquette', which itself comes from the old French 'label', meaning 'ribbon'. This usage has been attested since the seventeenth century and is based on the meaning of 'band, the fringe of something' derived from the words 'label, lambel, labeau and lambeau', in Old French. The label designates a piece of torn fabric, which by nasalization will become 'lambel' then 'lambeau' in the fifteenth century. Two aspects account for the use of the label in the Middle Ages: a decorative dimension that can be seen on clothing but, above all, an identification and symbolic function on the coats of arms of knights. From that time on, the 'lambel' thus affirmed important notions of visibility and identity, such as an emblem used in a family's coat of arms (Fournier 2014). After circulating for a few centuries in the Anglo-Saxon world, the word returned to France in 1899 with the English meaning of the label, as 'a strip of paper stuck on a product and giving indications such as its name, origin, composition or use'.[7] By transforming its meaning from 'ribbon' (lambel) to 'label' (label), the word evolves towards the idea of the sign by reassigning attributions of distinctiveness that return to its original meaning.

Nowadays, the definitions of the word 'label' are mainly based on an economic term from the business world. The Larousse dictionary[8] gives us the four meanings attributed to the term today: 'label or special brand created by a professional association', 'sign, the brand under which someone presents himself', 'synonymous with a label in the computer industry'[9] and finally 'record publishing company'. We choose to focus on the first two definitions. One refers to a product intended for sale, in which the label comes to certify the origin, guarantee quality and conformity to manufacturing standards. It is, therefore, a process created by a group of actors, in this case, a trade union, in order to stand out from other similar products. We can also notice the idea of belonging to a social group, which gives echo to the origins of the word in old French in the family coat of arms or blazon. Moreover, we note the importance given to the regulations with criteria of production methods that evoke a possible standardization of the product. The second meaning conferred on the term indicates identification with a network by the expression ' to present himself', while the word 'brand' refers to a visual identity such as a logo, sign or symbol. Once again, the filiation with the old French can be read in the notions of emblem and visibility. In that sense, the use of the word label makes a transition from the realm of discourse to the economy of the sign and thus of the image (Baudrillard 1972).

In Italian, Spanish and German, the word 'label' does not exist. It is translated as 'marchio' in Dante's language and is defined as 'a sign that is printed or applied on an object to distinguish and recognize it, but also to read its qualities and characteristics'.[10] As for the label, the aim is to identify an object visually, but with a commercial connotation explicitly stated in the term 'trademark'. It can be graphic, figurative, but above all original to distinguish the products of a company from other similar objects. This distinction between 'trademark' and 'label', absent in Italian, reveals an ambiguity of terms but also some differences between the two expressions. By its commercial function, the trademark reflects private use, whereas the label, by its membership in a network, reflects a collective vocation. As explained by researchers Olivier Aïm and Emmanuelle Lallement,

> the label does not belong to the same economy of visibility or the same economy of attention as the trademark: it is not used to distract, but to reassure. Its prestige belongs to history and continuity (...). A label should never surprise. It must confirm the good functioning, the good quality or the good reputation of an object, a practice, a site. (Aïm and Lallement 2016, 23)

The notion of a label, originated used in the business world, has therefore gradually moved into the realm of public action, employing so-called urban marketing methods and becoming a political tool (Devisme and Dumont 2006). This transformation reveals the idea of a shift from the notion of a label to the concept of labelling as a way of thinking and a process of action. Here the term 'label' is intended as a tool for thinking about the city and thus as a concept that has consequences in the urban production[11] in three fields: image, stakeholders and projects. Applied like a logo guaranteeing quality, the label becomes the 'icon' to stimulate the attractiveness of the destination. Some cities are thus banking on their history, the authenticity of traditional practices that still exist or the exceptional character of their architecture. Promoted in political speeches as an argument in favour of a participatory project approach, the label questions the groups of actors involved. Indeed, the circulation of models and the discovery of similar projects transform the types of cooperation between stakeholders. The rules derived from 'the experience of others' (Arab 2007) thus enable stakeholders to set up new management and project modes while controlling the associated risks. Similar to a contract mark, the label is similar to a sign of partnership within a network of people who share the same ambitions and transfer a certain number of 'good practices' between them. Within this reconfiguration of the stakeholders, in what way are the modes of the conception of the projects upset and modified by the ECoC label?

Matera, a city with a double international recognition

In 1993, the 'Sassi and the park of the Rupestrian Churches of Matera' were included in the UNESCO World Heritage List. On a perimeter of 1016 ha, the cultural site includes the troglodyte areas formed by the Sasso Barisano to the north and the Sasso Caveoso to the south, but also the Murgia park where nearly 150 churches or monasteries dug into rock are spread out.[12] The park presents places dating back to the Paleolithic with many archaeological sites such as the 'Grotta dei Pipistrelli'[13] or the sites of Murgia Timone, symbols of the Neolithic period. The inscription of Matera echoes certain exemplarity as a model on different scales: a model of integration on a landscape scale, a model of sustainable urban development and a model of the use of limited resources on an architectural scale with an economy of available materials.

Twenty years later, a new nomination is given to the city with the title of European Capital of Culture, awarded on 17 October 2014 under the title 'Matera 2019, Open Future'. The city of the Sassi stands out in a two-stage national competition[14] in which it wins over Siena, Perugia, Lecce, Cagliari and Ravenna. Its slogan, based on a digital and open future, won the support of the jury of experts[15] at the final audition. In the case of Matera, the selection panel emphasized the strategic analysis of Matera as a 'small and medium-sized European city'[16] aiming to position itself at the forefront of a movement based on digital technologies and learning. The importance given to new technologies as vectors of aggregation for populations usually excluded from cultural actions is considered as a 'forward-looking and innovative approach for an ECoC',[17] as the selection panel points out. The stated ambition is to break away from the large institutions and mobilize the greatest number of participants in order to generate a new relationship with culture. To achieve this, the programme is based on two pillar projects: the 'I-DEA',[18] a platform of archives open to the whole region, and the 'Open Design School', a laboratory for experimentation on design and urban spaces.

Open Future, which vision for the city?

In order to understand the city visions linked to the European Capital of Culture labelling process, the keywords included in Matera 2019 bid book have been analysed and identified.[19] The document was written in 2014, five years before the event. For the members of the Matera 2019 Committee, the aim is to convince the European Commission jury to award the prestigious title to the most audacious cultural programme, but also to the one that best meets the two main criteria: European dimension and citizen participation. During the first reading of the dossier, a certain number of anglicisms caught my attention. Expressions such as 'cluster', 'network' or 'capacity building' were recurrent, that's why systemic research of these English terms has been held to analyse the linguistic dimension. This method made it possible to identify the principle of competition that emanates from the labelling process. The idea of distinguishing itself from other cities is predominant by the notion of 'leader', which aims to make Matera a pioneer city in terms of digital innovation and also distinguish itself with this specificity from other competing cities. Through pilot projects, the writers of the dossier wish to show the assets that can guarantee the city a position and a major role in the south of Italy and more widely in Europe. The bid, therefore, seems to be a process of transformation marked by a 'before and after', with an accelerated time during the last phase of the competition. The idea of a quicker time frame at the end of the race is indicated by the term 'shortlist', which recurs many times in the text and is always indicated in italics.[20] It is most often used to refer to the 'entry into the shortlist', such as the passage from one state to another. In the same way that an industrialized product must go through evaluation, measurement and monitoring phases, the city goes through various validation stages to obtain the honorary title of European Capital of Culture.

The issue of innovation, which is very present in the dossier, reinforces the political discourse of the transformation of a territory, as evidenced by the sentence 'a process of change'.[21] This dynamic is more broadly part of an ambition toward the 'Mezzogiorno'[22] to give it a new narrative, a 'new vision of the South'[23] in a region with limited accessibility, both in terms of physical and virtual infrastructure. With its slogan 'Open Future', the Matera-Basilicata 2019 Foundation chooses to emphasize the notion of openness. But who is this slogan for and what kind of openness is it about? As exposed in the bid book, two notions and one project characterize this concept: 'open culture, open data' and the project of Open Design School. The expression 'open culture' is the notion that generates this new way of thinking, while the second corresponds to the data collected and produced by this system. Finally, the Open Design School is the place that represents and tests the potential of the digital engagement system. The openness promoted by the Foundation is, therefore above all, technological and coupled with tools related to the digital world. It is also evidenced by the importance given to the 'Creative Commons'[24] license requested both in the projects resulting from the Foundation's cultural programme and in the urban actions undertaken by the City Council of Matera. The adjective 'open'[25] is thus combined with different expressions: the culture linked to this new way of thinking, the technological dimension whose specific tools are articulated around digital fields: 'digital media', 'digital storytelling', the collection of new data qualified as open data but also a political approach with an alternative mode of government indicated as 'Open Government'.

Another dimension recurrently appears in the dossier: the notion of 'cluster'. Most of the time used to designate networks of companies or small industries, and the expression is here understood as a group of cultural projects linked to the same theme. By using this word, the reference to the company is clearly marked, as is the idea of specialized sectors anchored locally but intended for a more global market. Emblem of the application, this notion of openness aims at engaging partnerships with other European countries and is combined with the idea of networks necessary to obtain the ECoC label. The keyword 'network'[26] is also strongly connected to European networks at the cultural, artistic and institutional levels. More nuanced, the expression 'partner' appeals both to local media (newspapers and television) and to European cultural operators but has a more local dimension by insisting on regional partners: municipalities of Basilicata and cultural associations. The presence of private actors is also revealed in this expression of 'partnership', which evokes funding strategies in the form of sponsors or community fundraising[27] for the inhabitants of Basilicata but living outside the region.

This participation in relation to the citizens and inhabitants of Matera is at the heart of the cultural programme with the expression 'capacity building', the main theme of the dossier and mostly written in italics. With this chosen typography, the hypothesis of a form of international concept adopted by the Foundation Matera-Basilicata 2019 can be put forward. The theory of 'capacity building' – intended as community capacity building – often refers to improving the skills of people in small companies so that they can achieve their goals and overcome their difficulties. Used by NGOs and governments, this method aims to guide and structure the development of certain territories and populations in crisis. The metaphor of building takes up the idea of creating skills step by step. The allusion to the building is paradoxical, however, since the Foundation's idea in Matera is to avoid building new infrastructures but to invest in intellectual resources. This idea of construction (from the verb 'to build') is a necessary progression to be made with human capital that cultural experts are aiming to form through learning partnerships. This is how the programme entitled 'Build Up' is constructed, aiming to raise the skills of the territory's political actors and cultural operators. The Foundation's discourse promotes learning through seminars where several international good practices are presented under the name of 'learning partnership'. The expression in italics in the dossier is used to refer to a programme of visits to five European cities for some public administration actors. Considered as a real political tool, the programme is stated as a remedy to go beyond the 'assistance'[28] logic of the southern Italian regions and propose a new management model for these marginalized areas. Nevertheless, this new form of assistance can be discussed, as it is no longer financial but moral and depending on international cultural experts who contribute to transfer ways of doing, thinking and imagining the city.

Attracting a 'creative class'?

The company's image is reflected in the construction of the application, particularly in the management of the process by the key figure of the Matera-Basilicata 2019 Foundation. The Foundation presents itself as the 'manager' of a structure aimed at implementing a project or even a product to be marketed, as evidenced by the expressions 'marketing'[29] and 'merchandising'.[30] By presenting itself as an entity detached from the power structures, the Foundation underlines its capacity to benefit from a strong autonomy, guaranteed according to these same actors, by the transparency of the decision-making processes. Through this great independence in the management of the process, the organization seems to be the 'manager' of the operation, mandated by political actors. Therefore, these political actors delegate the strategic planning of the city and region to the organization.

By going beyond the concept of the creative city, the authors of the Matera 2019 project are following the concept of the 'creative class'. Using this strategy, they give priority to a human capital considered as 'creative', and leave aside the topic of spatial amenities or technological progress. The importance is therefore given to the digital world, culture at the heart of the programme, the idea of learning and the managerial governance of the process. Beyond the artists, designers and architects, the 'creative class' imagined by the writers of the 'Open Future' dossier is aimed at the whole population but, above all, at the political actors, capable of making decisions. In this redistribution of roles, new professions are emerging and are based on the important communication strategy implemented for Matera 2019. Consequently, the 'Matera ChangeMakers',[31] are described as persons whose mission is to support cultural projects from a logistical point of view in 2019 and the 'Matera Links',[32] as people in charge of communicating citizen initiatives, particularly through social networks. Other professions are privileged such as digital expertise: 'hacker',[33] 'blogger',[34] 'social media strategist' or 'audience developer'.[35] In this plurality of very specific professions, the term 'team'[36] appears several times with teams dedicated to different sectors: the Changemakers team, the Matera Links team, the web team, the team working within the Matera 2019 committee, the team in charge of thematic cultural projects and finally the artistic direction team.

Structured according to project poles, the Foundation presents itself as a company capable of providing services and benefits for the city. In an injunction to participation – an official criterion to obtain the ECoC label – the Foundation considers the inhabitants as resources for culture, qualifying them as 'cultural inhabitants'.[37] However, the character of this participation remains poorly defined in the dossier, except for the experimental dimension to be implemented in these initiatives. The expression 'target group'[38] – used in the field of marketing to describe a group of buyers that one seeks to conquer or retain – which refers here to the inhabitants, leaves one wondering about the real dynamics of participation.

Conclusion

The analysis of the vocabulary used in the Matera 2019 application highlights the city models underlying in the cultural programme. Through the use of an entrepreneurial vocabulary, the stakeholders of the Matera-Basilicata 2019 Foundation propose to transform the classic modes of bureaucratic management and the way of conceiving urban design. Spatial and material transformations on the territory are finally left out of this reflection or refer to the developments that can usually be observed in 'smart cities'. The proposed solutions turn to travel practises ('bike sharing'[39] and 'car pooling') rather than to the creation of heavy infrastructure (rail network, road and motorways). This city, designed on the model of contemporary metropolises, contrasts with the image given to visitors, namely a timeless city with a unique architecture that offers a disconnection from unbridled urban rhythms. Moreover, the 'creative class', a characteristic of large metropolises, questions its adaptation to non-urban contexts and, more specifically, to medium-sized cities such as Matera. In agglomerations marked by a lack of infrastructure, high unemployment and massive emigration of young graduates, can the theory of the 'creative class' be sufficient? What about the population that does not have the language codes of this thinking or the artistic and creative talents? As Marc

Levine explains using the example of Baltimore, some cities need above all solutions to rebuild public schools rather than favouring the leisure society in order to attract the 'creative class' (Levine 2004). By relying primarily on a hypothetical so-called creative population, the values of social and spatial justice are likely to disappear. Translating this model into the field of public policy can therefore reinforce inequalities within urban spaces and have harmful consequences in terms of gentrification but also socio-spatial fragmentation.

Concerning the transformation of the urban image, the analysis of the vocabulary reveals a new vision for Matera, which aims to internationalize the city beyond the heritage and traditions it holds. In this perspective, local identity is reinterpreted to correspond to the vision of a 'creative city' represented through the mobilizations of concepts, imaginaries and discourses of techno-cultural futures. In parallel to the over-exposure of the Sassi, a forward-looking look is being developed, as evidenced by the title of the bid book: 'Open Future'. By using this slogan, the Matera 2019 committee aims to break away from the previous image of the Sassi and propose a new dynamic through the cultural label. The European Capital of Culture is thus understood as a 'laboratory' – an expression repeated several times by the director of the Foundation Matera-Basilicata 2019 – which reveals the desire to renew the image of the troglodyte city. In this vision of a creative and inventive city, the material and aesthetic dimension of the Sassi's heritage is reinvested in a dynamic of entrepreneurial development. The area is becoming a place synonymous with the creation and a source of inspiration, as the mayor of Matera explains: ' the site where the University stands right now, will become a Digital Hub center, where start-ups and companies that want to activate this process of innovation and creativity, will go'.[40] The fascinating and intriguing atmosphere of the Sassi, depicted by photographers during the twentieth century, is no longer seen merely as a picturesque landscape to be sublimated but as a source of economic investment to complement tourism. Drawing inspiration from international cultural policies, the ECoC label is therefore employed as a way of changing urban image and as a lever for transforming the territory thanks to new global visibility.

Notes

1. In the rest of the article I will use the acronym ECoC to refer to the European Capitals of Culture.
2. According to the American economist, creativity is today a driving force of contemporary capitalism, which is giving rise to a new social class: the creative class. (Vivant 2006). His theory is based on three factors that, according to him, are correlated with economic growth: "Tolerance, Technology and Talent" (Darchen and Tremblay 2008).
3. From 1985 to 2005, the "European Cities of Culture" – which will then become the "European Capitals of Culture" – are chosen by the government of the country designated to host the event.
4. Athens in 1985, Florence in 1986, Amsterdam in 1987, Berlin in 1988 and Paris in 1989.
5. Written in 2014 the bid book is entitled "Open Future".
6. Six field surveys have been carried out since the beginning of the thesis with about a hundred interviews, a follow-up of urban projects and a historical analysis of the city through archive work.
7. Translation from 'Trésor de la langue Française informatisé' XIX et XXe siècle, 2017, atilf.atilf.fr (consulted on 15.01.2020).

8. https://www.larousse.fr/dictionnaires/francais/label/45761?q=label#45699 (consulted on 15.01.2020).
9. Label in this case refers to a group of characters that can be used to identify and describe an item, file or package of information in computing.
10. www.treccani.it (Translation of "marchio" by the author).
11. The expression is here intended as the urban economy specialisation.
12. Website Unesco. https://whc.unesco.org/fr/list/670 (consulted on 1.04.19).
13. Literally "bat cave" (author's translation).
14. The competition takes place in two phases: pre-selection and final selection. On 20 November 2012, the Ministry of Culture launched a call for applications to all Italian cities. The European Parliament delegates to the Ministries of the State chosen to host the Cultural year the organization of the competition for the title of European Capital of Culture. On 23 April 2013, the Italian Minister of Culture signs the regulations for the "Competition for the title of Culture 2019 in Italy", which he published on the Ministry's website. By the deadline of 20 September 2013, 21 applications had been submitted: Aosta; Bergamo; Cagliari; Caserta; the cities of the Val di Diano with the Cilento Park; Erice; Grosseto-Maremma; L'Aquila; Lecce; Mantova; Matera; Palermo; Perugia, Pisa; Ravenna; Reggio Calabria; Siena; Siracusa; Taranto; Urbino; Venice. The pre-selection meeting took place in Rome from 11 to 15 November 2013. The jury then recommended six cities to participate in the final selection. (The Selection Panel's 2014. Selection of the European Capital of Culture in 2019 in Italy, Rome).
15. The Italian Ministry of Culture appointed the thirteen members of the selection panel on 21 October 2013. Seven are appointed by the European institutions (European Parliament, Council of the European Union, European Commission and Committee of the Regions) and six by the Ministry in consultation with the European Commission. (The Selection Panel's 2014. Selection of the European Capital of Culture in 2019 in Italy, Rome).
16. The Selection Panel's. 2014. Selection of the European Capital of Culture in 2019 in Italy, Rome.
17. The Selection Panel's. 2014. Selection of the European Capital of Culture in 2019 in Italy, Rome.
18. The Institute of Demo-Ethno-Anthropology (translation by the author).
19. The dossier is written in two versions: one Italian and one English. My analysis is based on the Italian version to identify the english expressions.
20. The term appears a total of 13 times in the document.
21. P.3 of the bid book.
22. The term "Mezzogiorno" refers to the regions of southern Italy. These are usually characterized by low economic development and lack of industrial infrastructure (Cersosimo and Donzelli 1996).
23. P.3 of the bid book.
24. The use of this license – created to have more flexible intellectual property rights – is indicated as an obligation for projects related to the Matera 2019 programme. The expression is indicated on pages 10, 11, 20, 37, 51, 63 and 82.
25. The adjective appears 155 times in the bid book.
26. The term appears 22 times throughout the bid book and describes the international networks, the links with the European institutions but also the social networks considered as spaces for debate.
27. The principle of "crowdfunding" is enunciated as a way of involving citizens who have emigrated from Basilicata. They are invited to participate through symbolic funding to support the event (12 and 98 of the bid book).
28. Page 9 of the English bid book: "Engage in a systemic capacity-building programme for socio-cultural operators and local authorities in Basilicata. At the heart of the project-building efforts of operators in the creative and cultural fields is a strong cooperative philosophy that supersedes the culture of welfare. Their efforts focus on enhancing networking

capacities, forming new sustainable models for cultural action.". In the italian bid book it is the term of "assistance" which is used.

29. The term is used to refer to local economic development (5, 88) with a budget allocated to the promotion of the tourist destination (94, 95) and an important strategy with sponsors (95).
30. The word is used to talk about the objects to be created and sold for the Matera 2019 event (98, 99).
31. These are actually project managers who accompany local cultural operators to implement their production in 2019 (show, exhibition, concert) (9, 30, 54).
32. People mobilized to create a link between the inhabitants of Matera and the tourists (9, 18, 30, 54).
33. The term is used to refer to potential participants in the Open Design School (53).
34. The term is used to refer to the potential participants in the Open Design School (53) but also to the project entitled Jazz Notes in which the bloggers are tasked with telling the story of the event.
35. This is the role of the actors of Matera Links who, through seminars and visits, will have to collect citizens' initiatives (54).
36. The expression appears nineteen times in the file.
37. Translation for "abitante culturale".
38. The expression is used on p.45 to refer to the cultural project aimed at a wide group of citizens: children, teenagers, old people, young people, students, disabled people, migrants, emigrants, prisoners, people with mental disorders … It is also used on p.54 as part of the mission assigned to Matera Links.
39. This mode of transport is described as "smarter, alternative and sustainable" (96, 101).
40. Interview with the mayor Raffaello De Ruggieri, 18 october 2017 in Matera (Translation by the author).

Disclosure statement

No potential conflict of interest was reported by the author(s).

References

Aïm, O., and E. Lallement. 2016. "Labels du patrimoine: une nouvelle dialectique entre art et économie?" *Juris Art* 41: 23–26.

AlSayyad, N. 2013. *Consuming Tradition, Manufacturing Heritage: Global Norms and Urban Forms in the Age of Tourism.* London: Routledge.

Ambrosino, C. 2019. "Ville créative et renaissance urbaine. Retour sur la genèse intellectuelle d'un modèle urbain du XXe siècle." *Riurba* 6.

Arab, N. 2007. "À quoi sert 'expérience des utres?." *Espaces et sociétés* 131 (4): 33–47. doi:10.3917/esp.131.0033

Ashworth, G., and H. Voogd. 1990. *Selling the city: Marketing approaches in public sector urban planning.* Cardiff: Belhaven Press.

Baudrillard, J. 1972. *Pour une économie politique du signe. Coll. Les Essais.* Paris: Gallimard.

Béal, V., R. Epstein, and G. Pinson. 2015. "La circulation croisée." *Gouvernement et action publique* 3: 103–127. doi:10.3917/gap.153.0103

Bianchini, F., and M. Parkinson. 1994. *Cultural Policy and Urban Regeneration: the West European Experience.* Manchester: Manchester University Press.

Bianchini, F., and M. Parkinson. 1995. *The Creative City.* Londres: Demos.

Bourdin, A. 2005. "La 'classe créative' existe-t-elle?" *Urbanisme* 344.

Bourdin, A., and J. Idt. 2016. *L'Urbanisme des Modèles. Références, Benchmarking et Bonnes Pratiques.* Paris: L'Aube.

Cersosimo, D., and C. Donzelli. 1996. "Mezzo giorno e mezzo no. Realtà, rappresentazioni e tendez del cambiamento meridionale". *Meridiana*, 26–27.

Coudroy de Lille, L., and O. & Ratouis. 2019. "Le langage de l'urbanisme. Enjeux et limites d'une spécialisation." In *Les mots des urbanistes. Entre parlers techniques et langue commune*, edited by C. L. de Lille, and O. Ratouis, 235–279. Paris: L'Harmattan.

Darchen, S., and D. Tremblay. 2008. "La thèse de la « classe créative »: son incidence sur l'analyse des facteurs d'attraction et de la compétitivité urbaine." *Interventions économiques* 37.

Denuit, R. 2018. *Capitales européennes de la culture: un rêve de Melina*. Nivelles: Académie royale de Belgique.

Devisme, L., and M. Dumont. 2006. *Les métamorphoses du marketing urbain*. EspacesTemps.net, posted online on February 4.

Epstein, R. 2013. "Les trophées de la gouvernance urbaine." *Pouvoirs locaux* 97: 13–18.

Evans, G. 2001. "Cultural Planning: An Urban Renaissance?" *Town Planning Review* 73 (4): 489–490.

Fijalkow, Y. 2017. *Dire la ville c'est faire la ville: La performativité des discours sur l'espace urbain*. Villeneuve d'Ascq: Presses Universitaires du Septentrion.

Florida, R. 2002. *The Rise of the Creative Class, and how it is Transforming Work, Leisure, Community and Everyday Life*. New York: Basic Books.

Fournier, M. 2014. *Labellisation et mise en marque des territoires*. Clermont-Ferrand: Presses universitaires Blaise-Pascal.

Jones, Z. 2020. *Cultural Mega-Events: Opportunities and Risks for Heritage Cities*. London: Routledge.

Jones, Z., and D. Ponzini. 2018. "Mega-events and the Preservation of Urban Heritage: Literature Gaps, Potential Overlaps, and a Call for Further Research." *Journal of Planning Literature* 33 (4): 433–450. doi:10.1177/0885412218779603

Levine, M. (2004). La « classe créative » et la prospérité urbaine: mythes et réalités. Conference presented in Villes Régions Monde, INRS-Urbanisation, Culture et Société, Montreal, May 20.

Ponzini, D., and U. Rossi. 2010. "Becoming a Creative City: The Entrepreneurial Mayor, Network Politics and the Promise of an Urban Renaissance." *Urban Studies* 47 (5): 1037–1057. doi:10.1177/0042098009353073

Richards, G., and J. Wilson. 2004. "The Impact of Cultural Events on City Image: Rotterdam, Cultural Capital of Europe 2001." *Urban Studies* 41 (10): 1931–1951. doi:10.1080/0042098042000256323

Russo, A. 2002. "The 'Vicious Circle' of Tourism Development in Heritage Cities." *Annals of Tourism Research* 29 (1): 165–182. doi:10.1016/S0160-7383(01)00029-9

Vanolo, A. 2008. "The Image of the Creative City: Some Reflections on Urban Branding in Turin." *Cities* 25 (6): 370–382. doi:10.1016/j.cities.2008.08.001

Vivant, E. 2006. "La classe créative existe-t-elle ? Discussion de la thèse de Robert Florida." *Les Annales de la recherche urbaine* 101 (1): 154–161. doi:10.3406/aru.2006.2683

Zukin, S. 2009. *Naked City: The Death and Life of Authentic Urban Places*. Oxford: University Press.

Zukin, S. 2012. "Competitive Globalization and Urban Change." *Rethinking Global Urbanism*, 17–22.

Heritage and cultural mega-events: backgrounds, approaches and challenges

Jacek Purchla

ABSTRACT
Cultural heritage is not simply the stock of a given culture's material assets, but also the sum of our memory and identity – and this is part of the increasing significance of cultural heritage. Cultural heritage today is not only a resource in need of protection, but also a potential that must be exploited for future development. The three issues of significance for the future condition of our historic resources should be emphasized. The first is the rapidly accelerating process of transformation of the entire built heritage stock. The second is the increasing need for heritage. The third issue is the clear shift in emphasis from the value of the heritage to the needs of society. This breeds an increasing need for us to redefine ourselves and open up to others, so promoting the interaction between social capital and heritage is becoming vital for the rebuilding of social cohesion. Social capital should be nurtured as an opportunity both for the protection of cultural heritage and for its exploitation as a pro-development resource. It is also a big challenge for the Cultural Mega-events that may, and should, be well melded into a policy of building a broad base of heritage awareness.

Cultural heritage is not simply the collective stock of a given culture's material assets. It also includes our memory and identity, making it vitally important for society. While cultural heritage still requires ongoing protection today, it also represents a great potential for future development. Heritage is people: its creators, interpreters, and users. Herein lies the key to discussing the dynamic process of creating and reinterpreting heritage, as well as of protecting and valorizing it at the urban scale. Recently, the planning and implementation of mega-events – cultural mega-events especially – have interacted more and more intensively with these processes. This article builds on Ponzini's introduction and previous reflections of the authors, paying specific attention to European heritage cities and to the European Capital of Culture programme to develop a commentary to the key questions that have emerged in the 'Cultural Mega-events and Heritage: Challenges for European Cities' special issue and its contributions.

Cultural heritage and its social dimensions

One can broadly understand cultural heritage as 'meaningful pasts that should be remembered' (Macdonald 2010, 1). In this sense, people create heritage and, as Graham, Ashworth, and Tunbridge (2000, 1) claim, the concept of heritage depends on 'almost any sort of intergenerational exchange or relationship, welcome or not, between societies as well as individuals'. People are the owners of cultural heritage and it is up to us, and not to a narrowly defined group of experts, to define its meaning and value. In this sense, as well for other good reasons, cultural heritage has been recognized as one of the basic human rights. Heritage emerges as a generator of different kinds of social activity and has potentially unlimited users. In recent years, the increasing contribution of intangible heritage to the overall GDP of cities has clearly induced a major shift in how society approaches heritage. An increase in heritage resources, their international exposure and accessibility is tantamount to the growth of the social and economic developmental potential of heritage (Jagodzińska et al. 2015). However, sustainability must be taken into consideration in every field of policy and there has been increasing importance placed on the need for long-term strategies and plans. Environmental responsibility has also long been widely acknowledged as a collective social obligation towards future generations, resulting in long-term visions focused on the protection of the cultural environment, as evident in ongoing cultural policy debates. Yet such efforts cannot frame social and political mobilization as merely straightforward or easy liaisons between social and cultural interests (Purchla 2012). Any investigation of the cultural, political and economic impact of heritage on contemporary society pushes one to consider critical questions regarding who owns heritage and why as well as the social functions of heritage (identity building, economic benefit, etc.). Inevitably, questions about the relationship between such broad-stroke pictures of heritage and cultural mega-events arise and, despite its academic and policy importance for Europe today, only a few contributions have broached the subject (Jones and Ponzini 2018).

Phases and urban implications of the European Capital of Culture programme

The integration of heritage and artistic mega-events has a long tradition in Europe. Edinburgh and Salzburg serve as classic examples of the success of locating mega art-festivals into the priceless heritage context of those historic metropolises (Purchla 1996). Today, the project of the European Capital of Culture is without a doubt the European Union's greatest success in the sphere of culture. Melina Mercouri's original initiative from the mid-1980s has undergone a telling evolution. Its initial romantic vision and intention to foster better mutual understanding among Europeans gradually became an important tool for urban and economic development by the early 1990s. The programme transformed into an increasingly well-defined project for the betterment of the cities of Europe. While in the first phase, the cities chosen for the title (mainly through political pressure) were the grand capitals of European civilization, such as Athens, Florence, Amsterdam, Berlin, and Paris, the 1990s brought a different approach to the fore. The nomination of Glasgow in 1991 was a breakthrough – for it represented a city of an entirely different calibre and significance than its predecessors and for Europe at large. Glasgow's difficult period of socio-economic transformation successfully exploited culture, activating the international exposure as a driver for change. From Antwerp in 1993 to Lille in 2004, there have been several instances of the ECoC (and host cities) following a similar strategy. The success of the ECoC project was so acclaimed in cities such as Glasgow, Antwerp and Bergen that Brussels decided to refine the programme even further. The rivalry among cities to win the bid at the national level became stronger and stronger throughout the 1990s and 2000s. The German case of Nuremberg losing to Weimar in the competition for the 1999 title is one decisive example. However, the selection process at that time was still mostly governed by political criteria rather than cultural or social. In this context, the European Council and Parliament selected nine co-host European Cities of Culture for 2000. The creation of the network of very different cities, from Reykjavik and Santiago de Compostela to Prague and Kraków and the political gesture of allowing several countries still only aspiring to EU membership to participate in the programme marked an important turn. Since the 2000s and increasing interest in cultural and creative economies (Florida 2002), city mayors along with average citizens began to understand the importance of culture in development. Following the largest ever wave of accession to the European Union, consisting of 10 countries, including Poland, the Commission's president, José Manuel Barroso, stated in 2004 that the EU had reached the stage in its history when its cultural dimension could no longer be ignored. Within this context, the true revolution arrived with the introduction of new, detailed criteria[1] in 2007. As the cases of Hull, Pafos and Wrocław in this special issue confirm, small- and medium-sized towns, often burdened with post-industrial ballast, aspired to radical changes through the transformational potential of culture and heritage to act as catalysts to bring about positive changes. They started using innovative cultural policy management strategies, creating new cultural infrastructure and following the past approach of the ECoC as a reference model (Purchla 2011).

The challenge of mega-events for historic cities and their landscapes

Historic cities (for example, those included in the UNESCO World Heritage List) are situated within a particular arena in the balance between heritage and cultural mega-events. 'World Heritage' designation is becoming more meaningful in terms of conservation management and protection, where environments from the past and heritage have an impact on present-day development. It is no coincidence that most cultural places on the UNESCO World Heritage List are urban sites. Today's urbanization is characterized by an unprecedented speed and scale of growth. This poses a particular challenge to cities' heritage, raising questions about its protection and use. Mega-events are clearly part of broader urban globalization processes and can be considered to present both significant problems and opportunities in this regard.

The rapid transformation of our cultural cityscape often leads to its degradation if the technical and policy approaches are not capable of anticipating or interpreting ongoing change. The transition from a passive, static thinking about heritage, understood as a pious approach to relics of the past, to a new urban conservation approach, requires the protection of heritage in the midst of current realities that requires fundamental reinterpretations of the potential of heritage. The conception of cultural landscape conservation is crucial for historic towns.

Sixty years ago, Kevin Lynch published his pioneering work *The Image of the City* (1960), presenting the now well-known classic concept of Urban Landscape. A year later, Gordon Cullen applied the related concept of Townscape in his book *The Concise Townscape* (1961), developing a movement of defenders of the urban landscape (townscape movement). Passing from dreams about an ideal city to reflections on the city as a cultural landscape has allowed urban space to be seen not only as something functional but also in artistic terms, as a work of art. Based on these and other foundations, the international UNESCO community has long recognized the need to develop a strategy for the effective protection of the most valuable historic urban sites. We can speak of real achievement in terms of the success of the World Heritage Convention, one of the few almost universally ratified normative instruments. However, the majority of the existing documents ('charters') were conceived as if they were to be cast in bronze and fail to anticipate either the dynamics of evolving heritage concepts or the speed of change in urban settings. While there are still and always will be evolving conceptualizations, it is now clearly understood that the essence of culture is development, but also that culture is a driver of development.

The current prevailing discourse on the conservation of historic cities is based on the concept of the Historic Urban Landscape (HUL). This is not a new category of protection, but a tool for managing urban historic resources during times of rapid changes. The 2005 Vienna Memorandum was the catalyst of international dialogue on HUL protection; 2005 also marked the beginning of UNESCO's official interest in the issue. In October 2005, the 15th session of the General Assembly of States Parties to the World Heritage Convention adopted the text of the Declaration on the Conservation of Historic Urban Landscape. In the autumn of 2009, the General Assembly of States Parties requested the World Heritage Centre to prepare a draft document on the HUL concept – open to international discussion by the Vienna Memorandum – for the session of the UNESCO General Conference. In November 2011, the UNESCO General Conference in Paris adopted the text of the Recommendation on the Historic Urban Landscape.

The idea of Historic Urban Landscape does not only imply a revision of the Washington Charter of 1987, but provokes a reflection on the complex issue of the conservation of historic cities. The management of heritage potential in historic towns today requires an interdisciplinary and integrated approach, not only from decision-makers but also from town inhabitants.

Cultural heritage is an important part of the urban environment and plays an important role in development. The relation with mega-events inevitably requires a consideration of local heritage's particular values and to define the rules of sustainable development.

European challenges for heritage: acceleration, needs for heritage and social demands

In conclusion, it is important to focus on a few of the most pressing challenges, three significant issues impacting the future condition of our historic resources. They clearly connect to the ways cultural mega-events exert pressure on heritage; however, they also expand beyond and respond to more general issues as well.

First, the process of transformation of the built heritage stock is rapidly accelerating, a consequence of the increasing wealth of cities and societies, changes in the usage of space and the liberalization of land-use control, global tourism trends as well as induced cultural transformations. These factors tend to surpass the traditional way in which historic monuments used to be identified and protected (sites that, for whatever reason, survived long enough to be recognized as monuments). Nowadays, the importance of an active conservation policy – pursued with an understanding of the nature of the changes that are occurring – is crucial for nurturing urban transformation that aligns with the meanings attributed to a place by different actors, stakeholders and communities. Mega-events and cultural mega-events may use the places and symbols of historic cities in different ways, from serving as the backdrop of celebrations (I mentioned the first generation of ECoCs, but consider also the case of the Liverpool 2008 ECoC as presented in this issue by West) to a more intertwined understanding of people's use of place (the case of the Pafos 2017 ECoC, as explained by Dova et al. in this issue). Only the sites that have attributed a monumental value or that which can be charged with further cultural value (in whatever sense) by mega-events and other changes will have any chance of ongoing preservation and protection from destruction or severe alteration. This is why heritage policy actors should not shy away from understanding and contributing to change and drive the collective process of heritage selection and sense-making (as attempted by the city of Hull in its 2017 UKCoC programme regarding its maritime past and by the Wrocław 2016 ECoC programme with its wide array of community heritage at all levels, scales and in multiple locations).

The second challenge is the increasing interest in and need for heritage – in all its forms, above all architectural and urban monuments – in contemporary European cities. This is a logical consequence of the previous point and a necessary reaction to the all-standardizing forces of mass tourism and cultural globalization more generally. The potential to exploit heritage in tourism is proportional to its value but also to its legibility, authenticity, accessibility by tourists, visual attractiveness and other factors that may be hampered by growing masses of visitors in a short period of times as can occur with mega-events. This 'demand for heritage' will influence the character of preservation of complexes and urban landscapes, putting notions of heritage into traction with mass culture, tourism and consumption. This challenge is of crucial significance. The monument-as-document, is valued, above all, as a unique work. Meanwhile, a cultural product – and the monument-as-product – may also be appreciated for its 'novelty' value, even though this is in opposition to its 'antique' value as expressed by Alois Riegl (1903). The 'completeness' and 'legibility' facilitate an understanding of the monument that tend to become more important than the authenticity of its form and substance. The extent to which the monument-as-product is adapted to serve all the functions expected by the user (in most part, the tourist) is a critical matter that may be connected to attractiveness, to the appeal and acceptance by consumers. In the case of the monument-as-document-as-artwork, such elements are considered of secondary importance and should have no impact on the methods of preservation. While a static understanding of authenticity ceases to be the dominant theory, there is still room for evolution in the thinking of urban landscapes as they interact with new and emerging uses of the urban environment, including a well-intended interpretation of sustainable tourism (the crisis of the previous model and the risks of alternatives that are primarily based on narration are evident in the case of the Matera-Basilicata 2019 ECoC, as discussed in this special issue). Of course, putting the last issue into practice is a great challenge for many European cities and the stresses brought by a mega-event presents an important moment to test for the long term, as demonstrated by the case of Genoa 2004 ECoC in this special issue.

The third challenge is the clear shift in emphasis from the value of heritage itself to the multiple and sometimes contrasting needs of society and contemporary cities. Heritage has enjoyed a special status in cities, as an element of the past that required separate treatment in terms of planning, finance and use. This special treatment of monuments in particular depended on the belief that heritage is not the property of contemporary generations. The latter are not supposed to freely use heritage without considering future generations. In recent decades, the reality of urban policy-making shows that the assumption that we as the sole proprietors of heritage has spread beyond expectations (Böhm et al. 2009). Mega-events may be moments to perceive historic city transformation in a long-term perspective, but in many, cases decision-makers are more focused on meeting the hard deadline for the celebrations.

In conclusion, it is vital to emphasize that our identity today is experiencing a rapid evolution from a defensive outlook to an innovative stance. This breeds an increasing need for us to redefine ourselves while also opening up to other ideas to promote the interaction between social capital and heritage as vital components in rebuilding social cohesion.

In order to face these challenges, it is crucial that the public sector adopts new strategies for developing social capital respecting cultural heritage. As part of planning processes, new cultural policy arrangements and the delivery of mega-events, social capital should be nurtured as an opportunity both for the protection of cultural heritage and for its exploitation as a pro-development resource. This thinking can apply to both tangible and intangible heritage, the latter rapidly gaining in significance. Cultural Mega-events may, and perhaps also should, be well integrated into policies that build a broad base of heritage awareness to work towards these goals (Ponzini et al. 2020). In this sense, an integrated approach must be developed, both by mega-event and heritage actors, as the *Charter for Mega-events in Heritage-Rich Cities*[2] clearly advocates. More research and public debate on the matters of this *European Planning Studies* special issue will help in deepening these points and spreading the findings at all levels, in Europe and beyond.

Notes

1. Pursuant to Decision 1622/2006/EC of the European Parliament and of the Council of 24 October 2006 establishing a Community action for the European Capital of Culture event for the years 2007–2019, one of the fundamental criteria for selection of the ECC city is a European dimension to the proposed actions; the other criteria is 'the City and Citizens', which is intended to foster involvement of the city's residents in preparing and conducting the projects. Another important tenet was incorporation of culture into the candidate cities' development strategy.

2. See this webpage for further information regarding the Charter: https://mck.krakow.pl/homee-charter.

Disclosure statement

No potential conflict of interest was reported by the author(s).

References

Barroso, J. 2004. Europe and Culture, speech delivered in Berlin on 26 November 2004 during the conference A Soul for Europe.

Böhm, A., P. Dobosz, P. Jaskanis, J. Purchla, and B. Szmygin, eds. 2009. *Raport o systemie ochrony dziedzictwa kulturowego w Polsce po roku 1989*. https://www.nck.pl/badania/raporty/system-ochrony-dziedzictwa-kulturowego-w-polsce-

Cullen, G. 1961. *The Concise Townscape*. London: The Architectural Press.

Florida, R. 2002. *The Rise of the Creative Class*. New York: Basic Books.

Graham, B., G. Ashworth, and J. Tunbridge. 2000. *A Geography of Heritage: Power, Culture, and Economy*. London: Routledge.

Jagodzińska, K., J. Sanetra-Szeliga, J. Purchla, C. Van Balen, C. Thys, A. Vandesande, and S. Van der Auwera. 2015. *Cultural Heritage Counts for Europe: Full Report*. Krakow: International Cultural Centre.

Jones, Z., and D. Ponzini. 2018. "Mega-events and the Preservation of Urban Heritage: Literature Gaps, Potential Overlaps, and a Call for Further Research." *Journal of Planning Literature* 33 (4): 433–450. doi:10.1177/0885412218779603

Lynch, K. 1960. *The Image of the City*. Cambridge, MA: MIT Press.

Macdonald, S. 2010. *Difficult Heritage: Negotiating the Nazi Past in Nuremberg and Beyond*. London: Routledge.

Ponzini, D., F. Bianchini, N. Tzortzi, and J. Sanetra-Szeliga. 2020. *Mega-Events and Heritage: The Experience of Five European Cities*. Krakow: Inetrnational Cultural Centre.

Purchla, J., ed. 1996. *The Historic Metropolis: A Hidden Potential?* Krakow: International Cultural Centre.

Purchla, J., ed. 2011. "Laboratory of Change." *Herito* 5: 110–123.

Purchla, J. 2012. "Cultural Heritage an Social Capital." In *VVAA. The 1st Heritage Forum of Central Europe*, edited by J. Purchla, 69–77. Krakow: International Cultural Centre.

Riegl, A. 1903. *The Modern Cult of Monuments: Its Character and Its Origin*. Cambridge, MA: MIT Press. 1982 edition.

Index

Note: Page numbers followed by "n" denote endnotes.

For Product Safety Concerns and Information please contact our EU representative GPSR@taylorandfrancis.com
Taylor & Francis Verlag GmbH, Kaufingerstraße 24, 80331 München, Germany

www.ingramcontent.com/pod-product-compliance
Ingram Content Group UK Ltd.
Pitfield, Milton Keynes, MK11 3LW, UK
UKHW052053230926
475702UK00025B/1149
* 9 7 8 1 0 3 2 4 4 5 8 3 0 *